T0331511

Occupational Injury

Occupational Injury: Risk, Prevention and Intervention

EDITED BY

ANNE-MARIE FEYER
National Centre of Occupational and Environmental Health Research,
New Zealand

AND

ANN WILLIAMSON
School of Psychology, University of New South Wales

Taylor & Francis
Publishers since 1798

UK Taylor & Francis Ltd, 1 Gunpowder Square, London EC4A 3DE
USA Taylor & Francis Inc., 1900 Frost Road, Suite 101, Bristol, PA 19007

British Library Cataloguing in Publication Data

A catalogue record for this book is available from the British Library

ISBN 0-7484-0646-8 (cased)
ISBN 0-7484-0647-6 (paperback)

Library of Congress Cataloging Publication Data, are available

Cover design by Hyberts Design & Type
Computerset in Times 10/12pt by Best-set Typesetter Ltd., Hong Kong
Printed in Great Britain by T. J. International, Padstow, UK

Contents

Foreword

Over eight decades, MMI Insurance has grown to be Australia's largest workers' compensation insurer, providing coverage to approximately one in five working Australians. Since MMI opened its first medical clinic in 1932, it has recognised the intrinsic link between injury and claims management. It is appropriate, therefore that MMI was a major sponsor of the Symposium on which this book was based. Workplace health and safety is a serious economic and social issue in Australia as it is in many countries. Based on figures provided by Worksafe Australia, over 500 traumatic deaths occur at work every year and at any one time, about 200,000 people cannot work at all due to occupational injury or disease. The total cost of work-related injury and disease is conservatively estimated to be around $20 billion per year.

All Australians bear this cost burden in some way, so clearly we need to work together to reduce it. Part of the approach we have taken has been to work on forging better links and integration between our injury compensation, return to work and prevention activities with government, other insurers and business. In places, such as in a number of states of Australia, where tariff pricing is imposed rather than market-driven, insurers can only compete on service delivery. This can involve not only delivery of effective claims management, but also proactive case and portfolio management to identify opportunities for early return to work, hazard reduction and improved safety management practices. Conferences such as this, that bring together expertise from around the world, help to promote new information and the results of research. This is vital for continuing improvement in safety management in the insurance industry, health and safety professions and government.

Mr Pieter Franzen
General Manager
MMI Insurance

Preface

Occupational injury is a major potentially preventable health problem in working communities. Even so, too many workers are injured or killed at work. For example, mean annual rates of fatal injury while working for persons in the employed civilian labour force during 1982–84 were 5.9/100 000 in the US and 6.7/100 000 in Australia (Stout *et al.*, 1990). Although fatalities provide relatively unambiguous indicators of the extent of the problem at the most severe end of the spectrum, they are, of course, only part of the picture of work injury. Recent estimates from the US indicate that traumatic occupational fatalities number 8000 to 10 000 per year (Saruda and Emmett, 1988; National Institute for Occupational Safety and Health, 1986); probably approximately 10 million non-fatal traumatic occupational injuries occur each year. In Australia, the most recent statistics indicate that approximately 108 000 compensated cases of occupational injury and poisoning were recorded (National Occupational Health and Safety Commission, 1996). The number of less severe injuries not covered by these statistics (those requiring up to five days off work) is likely to be several times this number.

In February 1996 the Third International Conference on Injury Prevention and Control was held in Australia. Associated with that meeting, an international Occupational Injury Symposium was held in Sydney under the joint sponsorship of the National Occupational Health and Safety Commission (Worksafe Australia) and the MMI Insurance Group, one of the major workers' compensation insurers in Australia. The symposium provided the opportunity to bring together leading researchers and professionals actively working in the field as plenary and keynote speakers to participate in sessions arranged around a number of major issues in occupational injury and safety. This book contains a selection of the invited papers from the symposium.

The chapters in this book reflect the views of a wide-ranging group of authors at the cutting edge of the field. In broad terms, the topics covered include estimating the size of the problem (different analytical techniques and their utility), the nature of the causal agents (e.g. the role of behaviour, the role of risk, the role of organizational processes), issues in prevention of injury (e.g. strategies for workplace interventions, strategies for evaluation, the role of enforcement) and the nature of the outcomes of occupational injury (e.g. the role played by compensation processes). Thus, the topics included in the book are rather diverse and provide a multifaceted view of occupational injury, but all converge on one of the fundamental aspects of the problem. The chapters are grouped around major topic areas with a brief introduction to each part that anchors the chapters to the central theme and highlights cross-over to other topic areas.

Together, the contribution of the plenary and keynote speakers provide an integrated picture of some of the major issues in occupational injury and safety. This book ensures that the information survives beyond the symposium and documents some of major themes in the field at the end of the twentieth century.

References

NATIONAL INSTITUTE FOR OCCUPATIONAL SAFETY AND HEALTH (1986) A proposed national strategy for the prevention of severe occupational traumatic injuries. In *Proposed national strategy for the prevention of leading work-related diseases and injuries*, Atlanta: National Institute of Occupational Safety and Health.

NATIONAL OCCUPATIONAL HEALTH AND SAFETY COMMISSION (1996) *Compendium of Workers' Compensation Statistics, Australia, 1994–95*, Canberra: Australian Government Printing Service.

SARUDA, A. and EMMETT, E.A. (1988) Counting recognized occupational deaths in the United States. *Journal of Occupational Medicine*, **30**, 868–72.

STOUT, N., FROMMER, M.S. and HARRISON, J. (1990) Comparison of work-related fatality surveillance in the USA and Australia. *Journal of Occupational Accidents*, **13**, 195–211.

Contributors

François Béland
Groupe de recherche
  interdisciplinaire en santé
Faculté de médicine sociale et
  préventive
Université de Montréal
C.P. 6128 Succ. Centre-Ville
Montreal
Quebec H3C 3J7
Canada

Yossi Berger
National Occupational Health and
  Safety Unit
Australian Workers' Union
685 Spencer Street
West Melbourne
Victoria 3003
Australia

Verna Blewett
New Horizon Consulting Pty Ltd
13A Dudley Road
Marryatville
South Australia 5064
Australia

Sally A. Brinkman
Environmental Health Branch
South Australian Health Commission
PO Box 6
Rundle Mall
Adelaide
South Australia 5000
Australia

Robert Cox
7 Margaret Street
Glenalta
South Australia 5052
Australia

Nicole Dedobbeleer
Groupe de recherche
  interdisciplinaire en santé
Faculté de médicine sociale et
  préventive
Université de Montréal
C.P. 6128 Succ. Centre-Ville
Montréal
Québec H3C 3JC
Canada

David M. DeJoy
School of Health and Human
  Performance
Department of Health
  Promotion and Behavior
Stegeman Hall
University of Georgia
Athens
Georgia 30602-3422
USA

Tim R. Driscoll
National Institute of Occupational
  Health and Safety
  (Worksafe Australia)
GPO Box 58
Sydney 2001
Australia

Paul Freestone
Freestone Transport Pty Ltd
20 Barrie Road
Tullamarine
Victoria 3100
Australia

Robyn R.M. Gershon
Department of Environmental Health
  Science
Johns Hopkins University
Baltimore
MD 21205-2179
USA

Richard T. Gun
Department of Public Health
University of Adelaide
South Australia 5005
Australia

Andrew R. Hale
Safety Science Group
Faculty of Technology and Society
Technische Universiteit Delft
Kanaalweg 2b
2628 EB Delft
The Netherlands

Peter A. Hancock
Human Factors Research Laboratory
University of Minnesota
141 Mariucci Arena – Operations
1901 Fourth Street SE
Minneapolis
MN 55455
USA

Andrew Hopkins
Department of Sociology
Australian National University
GPO Box 4
Canberra
ACT 2601
Australia

Jan Hovden
Department of Organisation and
  Work Sciences
Norwegian Institute of Technology
Alfred Getz vei 1,
N-7034 Trondheim
Norway

Kenneth R. Laughery
Department of Psychology
Rice University
Houston
Texas 77005-1892
USA

James Leigh
National Institute of Occupational
  Health and Safety
  (Worksafe Australia)
GPO Box 58
Sydney
NSW 2001
Australia

Petra Macaskill
Department of Public Health and
  Community Medicine
University of Sydney
NSW 2006
Australia

Garry Mahon
Road Use Management and Safety
Queensland Transport
PO Box 673
Fortitude Valley
Queensland 4030
Australia

Michael P. Manser
Human Factors Research Laboratory
University of Minnesota
141 Mariucci Arena – Operations
1901 Fourth Street SE
Minneapolis
MN 55455
USA

Ewa Menckel
National Institute for Working Life
S-171 84 Solna
Sweden

Thomas Mitchell
Victorian Institute of Occupational
   Safety and Health Australia
University of Ballarat
PO Box 663
Ballarat Victoria 3353

Lawrence R. Murphy
Applied Psychology and Ergonomics
   Branch
National Institute for Occupational
   Safety and Health
Cincinnati
OH 45226
USA

Claire M. Pollock
School of Psychology
Curtin University of Technology
GPO Box U1987
Perth
Western Australia 6001
Australia

Denis Roberston
Road Transport Forum
GPO Box 1879
Canberra
ACT 2601
Australia

Jorma Saari
Finnish Institute of Occupational
   Health
Topeliuksenkatu 41 A
FIN-00250
Helsinki
Finland

John Sargaison
Queensland Mining Council
7th Floor
AGL House
60 Edward Street
Brisbane Queensland 4000

Harry S. Shannon
Institute for Work and Health
Suite 702
250 Bloor St East
Toronto
Ontario M4W 1E6
Canada

Andrea Shaw
Shaw Idea Pty Ltd
RSD E1452
Ballarat
Victoria 3352
Australia

Laurie Stiller
New South Wales Minerals Council
PO Box A244
Sydney South
NSW 2000
Australia

Nancy Stout
Division of Safety Research
Centres for Disease Control and
   Prevention, NIOSH
1095 Willowdale Road
Morgantown
West Virginia 26505-2888
USA

Kent P. Vaubel
Department of Psychology
Rice University
Houston
Texas 77005-1892
USA

Willem A. Wagenaar
Section of Experimental and
   Theoretical Psychology
Faculty of Social Science
Leiden University
Leiden
The Netherlands

Gerald J.S. Wilde
Department of Psychology
Queens University at Kingston
Ontario K7L 3N6
Canada

PART ONE

The data speak but what do they tell us?

INTRODUCTION

Accurate and reliable occupational injury data are considered to be the essential starting point for developing injury prevention programs. Such data tell us about who gets injured as well as how, when and where the injury occurred. These data therefore have the potential to identify targets for preventive efforts. Injury data also hold considerable potential as tools in the evaluation of progress towards achieving prevention goals.

Harnessing the potential that occupational injury data offer for assisting in prevention is clearly a key requirement in the field. Many of the methodological problems involved have been discussed in the literature (e.g. Pollack and Keimig, 1987; Driscoll, 1993). The fundamental expectations for injury data are that they should provide accurate information about the nature, extent and distribution of the problem. The chapters in this part consider aspects of using routinely collected data to describe the problem more effectively and to provide possible directions for preventive efforts.

The first chapter, by Macaskill and Driscoll, takes up the issue of examining options for improving and expanding the current framework for national occupational injury data obtained from routinely collected data. While the specific examples are drawn from data sources in Australia, the basic issues discussed in the chapter are generic. These authors conclude that even considering routinely collected data, the current framework could be extended. In particular the authors advocate linkage between routinely collected data sources because, individually, such sources only cover a subset of cases. With linkage, the unique strengths of various data sources could be combined to give more comprehensive coverage of the scope of the problem.

Better utilization of existing routinely collected data is also taken up in the chapter by Stout. This chapter describes the coding process employed to use narrative data obtained in surveillance systems more effectively. Such data are routinely included because of their potential to allow better understanding of the circumstances of injury and hence provide clearer directions for control of the hazards. Although most systems include collection of narrative data describing at least the immediate circumstances of injury, their analysis is generally perceived to be pro-hibitively time-consuming. Consequently, the narrative data are often not fully utilized. In this chapter Stout provides a method for using narrative data efficiently, as well as examples of application of the method. As Stout argues, the examples demonstrate the value of using narrative fields in surveillance systems to more specifically identify sources of injury.

The final chapter in this part provides the perspective of an end-user of occupa-tional injury data, i.e. those responsible for developing prevention programs. These authors take up the issue of one major limitation of routinely collected injury data, namely that these data often reflect practical collection concerns rather than con-cerns about what information is needed to support development of preventive strategies. They describe a project which attempted to use compensation-based data for identifying target areas for preventive efforts in the coal-mining industry in Australia. While the data were able to be manipulated to be more useful for highlighting possible directions for prevention, the project also highlighted some limitations for industry users of routinely collected injury data.

The potential of injury and accident data to play a more effective role in the development of preventive strategies is discussed elsewhere in this volume. Part Four takes up the issue of the emergence of a more systems-based approach to safety. A major implication of this for understanding the causes of injury is that the influences that need to be considered are rather more far-ranging than is obvious in the immediate circumstances of accidents and injury. While epidemiological de-scriptions of the nature, distribution and rate of work injury provide the important starting point for identifying risk of injury, by their very nature they provide only limited information about the wider circumstances of accidents. To be maximally useful in the process of developing prevention programs, occupational injury data need to be considered in the broader context of a system-based approach to safety in which injury data can provide much richer information about the causes of injury (Reason, 1990; Feyer and Williamson, 1991; Laflamme et al., 1991; Kletz, 1993; Chapter 10).

The chapters in this part take up the issue of how routinely collected data can be better utilized to provide information about the nature and extent of occupational injury, and information about where preventive efforts might be best targeted. Improvements to routinely collected sources of information and their use are both possible and desirable. The chapters in this part provide some insights into the directions such improvement might take.

References

DRISCOLL, T.R. (1993) Are work-related injuries more common than disease in the workplace? *Occupational Medicine*, **43**, 164–166.

FEYER, A.-M. and WILLIAMSON, A.M. (1991) A classification system for causes of occupa-tional accidents for use in preventive strategies. *Scandinavian Journal of Work, Environ-ment and Health*, **17**(5), 302–311.

KLETZ, T.A. (1993) Accident data – the need for a new look at the sort of data that are collected and analysed. *Safety Science*, **16**, 407–415.

LAFLAMME, L., DOOS, M. and BACKSTROM, T. (1991) Identifying accident patterns using the FAC and the HAC: their application to accidents at the workshops of an automobile and truck factory. *Safety Science*, **14**, 13–33.

POLLACK, E.S. and KEIMIG, D.G. (eds) (1987) *Counting injuries and illnesses in the workplace: proposals for a better system*, Washington DC: National Academy Press.

REASON, J. (1990) The contribution of latent failures to the breakdown of complex systems. *Philosophical Transactions of the Royal Society of London B*, **327**, 475–484.

CHAPTER ONE

National occupational injury statistics: what can the data tell us?

PETRA MACASKILL AND TIM R. DRISCOLL

1.1 BACKGROUND

A recent inquiry into occupational health and safety in Australia was very critical of current national reporting of occupational health and safety (OHS) statistics, which are, at present, largely based on successful workers' compensation claims for five or more days off work (Industry Commission, 1995). Major deficiencies were identified in the area of occupational disease data, particularly for diseases of long latency. Such criticisms are not unexpected given the well-known methodological difficulties associated with the study of disease. However, despite the much better data available for occupational injury, a number of important limitations were noted, including the lack of coverage of self-employed workers, the lack of data on less severe injuries, missing data, non-compliance with data standards, the restricted number of variables and limitations imposed by the coding used for those variables.

The criticisms outlined above raise questions as to what we require of national occupational injury statistics and how we can obtain data to meet those requirements. The aim of this chapter is to provide an overview of potential sources of routinely collected data for work-related injury in Australia and outline a number of methodological and practical issues associated with obtaining national statistics from these data. In examining the options for improving and expanding the current framework for national occupational injury data, we must address the issue of what we can realistically expect the data to tell us. Although this paper is based on the Australian experience, the issues are relevant to many countries.

1.2 INFORMATION NEEDS AND EXPECTATIONS

Information needs vary according to the area of interest and responsibility of the person or agency requiring the information. To a large extent these needs determine the expectations of what the data should be able to tell us. Although these expectations vary, an underlying requirement is that the data should provide an accurate and representative picture of occupational injuries across levels of severity ranging from traumatic fatalities to minor injuries. Common aims and expectations of data-collection systems for injury surveillance include:

- estimating the extent of the problem and monitoring injury rates over time;
- identifying patterns in injury rates;
- identifying groups to be targeted through prevention measures;
- aiding in the development of prevention measures;
- evaluating the effectiveness of prevention activities;
- identifying new or emerging hazards; and
- providing data in a timely fashion.

Occupational injury data provide a basis for comparing OHS performance across jurisdictions and industries at a point in time, and monitoring trends and changing differentials between groups over time. Accurate and reliable data are required to set goals for injury prevention and monitor progress towards achieving those goals.

1.3 POSSIBLE DATA SOURCES

The information needs outlined above can only be met through utilizing a range of data sources (Macaskill *et al.*, 1995). Common sources that are used in many countries for occupational injury surveillance and statistical reporting include death records, coroners' reports, hospital in-patient records, hospital emergency department data, workers' compensation claims, OHS agency investigation reports and surveys. This paper will focus on the first five of the above sources as these represent the major sources of routinely collected data that can contribute to OHS surveillance in Australia.

The likelihood of identifying an injury case through routine data collections increases with the severity of the injury, and hence routinely collected data often represent the major source of data for fatalities and serious non-fatal injuries. All fatalities are recorded in death registries and most serious non-fatal injuries are likely to come in contact with either the hospital system or the workers' compensation system (or both). Because the number of cases increases sharply as the severity of the injury decreases, survey methods provide a means of obtaining data on less severe but far more common work-related injuries.

In Australia, the eight state and territory governments have responsibility for most routine data-collection activities. This has led to inconsistencies in information systems between jurisdictions in terms of what items are collected and the methodology used. As a result, reaching agreement between the jurisdictions on data standards is an important prerequisite for pooling data at the national level.

1.4 DATA STANDARDS

For each data set, combining the data at a national level depends on common data items, definitions and coding. A common core or 'minimum data set' must be defined, with which the jurisdictions that collect the data are both willing and able to comply. In Australia, such agreements are in place for workers' compensation (National Occupational Health and Safety Commission, 1987) and also health-sector data (National Health Data Committee, 1995). Although these agreements provide a basis for compiling consistent national data, full compliance with these standards has not yet been attained. Maintaining compliance over time is also problematic. For example, the standard for the National Data Set for Compensation-based Statistics (NDS) specifies that information be recorded for all successful claims for five or more days off work. However, a number of states have recently moved to a threshold of 10 days' lost time. These changes reflect changing policies/priorities at the state level, which can have a major impact on national data collection activities. The difficulties associated with obtaining and also maintaining compliance with date standards are considerable.

By adopting data definitions and codes that are compatible with recognized national and international standards, it is possible to achieve comparability at a national and also international level. However, it is important to recognize that these data-coding standards are subject to change over time in response to factors such as changing categories of work and categories of injury. For instance, the International Classification of Diseases (ICD) is updated periodically, as are the Australian standard codes for occupation and industry as defined by the Australian Bureau of Statistics. Data codes reflect the state of knowledge and/or situation at the time they were developed. Not only does this mean that they must be reviewed and updated from time to time, but it also means that new or emerging issues may not be apparent in the data because of the constraints imposed by the coding scheme. The inclusion of a narrative field that allows for a brief written description can provide qualitative data that can be used in conjunction with the coded variables to obtain greater insight (Stout and Jenkins, 1995).

Where such changes to data-coding standards are a simple extension of the existing codes, they are not problematic. However, if the revised codes do not map onto the previous ones, problems arise in maintaining comparability over time. Concordance tables provide a method for redistributing codes between the 'old' and the 'new' classifications. However, the assumption that cases can be redistributed on a proportional basis is often not valid because the risk of injury may not be uniform for all persons in a given category and hence this approach can introduce bias.

The adoption of national data standards is clearly essential to provide a framework for obtaining consistent national data. This approach, however, carries the underlying assumption that national data should be seen as a subset or common core of data items collected at the jurisdictional level. Hence a national 'minimum data set' will usually be less detailed than the sources at the state level from which the data are drawn. Although critics point to the limited number of data items often available at the national level, it is clear that this is unavoidable. Nevertheless, it must be acknowledged that an important potential consequence of defining a national minimum data set is that the data items collected at the state level may be reduced to conform to the minimum standard or 'lowest common denominator'.

1.5 FATAL INJURIES

Given the importance of death as an outcome, virtually all deaths due to injury will be recorded. In Australia, not only are such deaths recorded in routine mortality statistics, but almost all cases are also investigated by a coroner. Although mortality data are computerized and are readily available at the national level, no indicator is provided that will clearly distinguish work-related injury deaths from other traumatic fatalities. Coroners' records do provide extensive information on all traumatic fatalities in sufficient detail in most cases to establish whether the death is work-related and provide details of the circumstances associated with the fatality. However, most of these records are not computerized and accessing them is time-consuming and expensive.

Despite these difficulties, coroners' files for the three-year period 1982–84 were used for the first work-related fatalities study conducted by Worksafe Australia. Detailed examination of paper records for all non-suicide deaths ($\approx$ 17000 cases over the three years) identified approximately 500 work-related traumatic fatalities per year for that period (Harrison *et al.*, 1989). Despite the fact that this figure is now somewhat out of date, it is still probably the most frequently quoted OHS statistic in Australia. It represents one of the few 'hard' numbers available. An important strength of this study was the extensive qualitative information available on most cases, which allowed for in-depth analyses in a number of areas based on one or more of the involved agencies (Mandryk *et al.*, 1996), industries (Erlich *et al.*, 1993), occupations (Driscoll *et al.*, 1994; Driscoll *et al.*, 1995), subjects (Corvalan *et al.*, 1994) or associated factors (Feyer and Williamson, 1991). The results of the study not only provided a relatively accurate assessment of the number of work-related traumatic fatalities but also provided valuable insights into causal factors.

Given the extensive resources required to undertake such a study, it is not surprising that a second study of work-related fatalities has only recently been initiated. The second study, which commenced in 1995, is adopting a methodology that uses a somewhat broader definition of work-relatedness but will nevertheless allow direct comparison with the previous study. National deaths data information covering all non-suicide traumatic fatalities occurring during the four-year interval 1989–92 were used to identify relevant coroners' files, of which there were more than 21000. The first year of the study has involved locating and examining the often extensive and relatively unstructured written records to ascertain which cases are work-related. The second year will be spent developing a computerized database and coding the information. Extensive narrative fields are included in the database to allow for qualitative as well as quantitative analysis of the data. These written descriptions also provide the possibility of re-coding variables in the future to allow comparison with other data sets that employ an incompatible coding scheme.

Given the importance of coroners' data to the general area of injury prevention, the coroners are developing a national coronial information system (Moller, 1994). One of the advantages of defining such an information system is that, when fully implemented, it should result in a more systematic approach to the investigation of work-related deaths. The planned computerized database will include coded variables and also extensive narrative descriptions, which will allow regular and more timely analyses of the data. The success of this project not only depends on the will and expertise of those involved but also on the required funds being made available.

In the long term, a computerized coronial information system should be an important component of a national OHS surveillance system. Although ready access to these data will be a major improvement on the current situation, it must be noted that the level of information available on most cases is unlikely to be as extensive as that obtained in the two work-related fatalities studies described earlier.

1.6 NON-FATAL INJURIES

Obtaining accurate, representative and detailed data on fatalities is far more straightforward than obtaining such data for non-fatal injuries. Although some routine data collections do capture information on the 'more severe' cases, defining severity is often difficult and a number of indicators are used. Given the wide range of non-fatal injuries that occur and the variation in severity of those injuries, it is clear that no single source of data can provide an overview of the extent of the problem. Also, most sources provide very limited information, if any, on factors associated with the injury. In practice, there is often a trade-off between the depth of information on each case and the coverage and representativeness of the data.

In Australia, at present, we rely largely on workers' compensation data to obtain statistics on work-related injury but do not utilize health-sector data to any great extent. The limitations and advantages of these data sources are discussed below.

1.6.1 Workers' compensation data

After examining what information was available at the time, a decision was made in 1987 to establish a national data set for compensation-based statistics to cover all successful claims for five or more days of lost time (National Occupational Health and Safety Commission, 1987). Although some states collect data below this threshold, adopting this cut-off provided a basis for compiling nationally comparable data. However, as noted earlier, some states have since moved to a higher threshold of 10 days' lost time.

A data standard and coding system was developed specifically for the NDS. Current data items include age, gender, occupation, industry, type of occurrence (nature of injury, bodily location, mechanism of injury and agency), duration of absence, extent of disability and payments. Allowance was also made for a short narrative description of the circumstances surrounding the incident. Subsequently, items relating to ethnicity and size of employer were also added. Standard codes were used for most variables, but a special coding system was developed for type of occurrence to reflect the types of injury that occur in the workplace. The first 'national' figures, using data from the jurisdictions that were able to supply the information in the required form, were published in 1994 and covered the period 1991–2 (Worksafe Australia, 1994). Results for 1992–3 are based on 160000 successful claims, of which approximately 80% are for injury (Worksafe Australia, 1995).

Implementing this standard and compiling a national data set has taken longer than expected. In fact, full compliance has not been achieved. Some data items have proved difficult for jurisdictions to collect. For instance, reliable information on the size of the employer is difficult to obtain and collecting data on ethnicity has proved to be problematic because this information is seen as 'sensitive' in some jurisdictions. Narrative descriptions are supplied by only one state. Compliance for other items is far better.

Although the limitations of compensation data were acknowledged in 1987, publication of the results has again drawn attention to these. In particular, the data do not cover all workers (e.g. the self-employed, who make up 15–20% of the paid work-force, are not included), less severe injuries that result in less than five days off work or fatalities that do not result in a claim by a dependent. Other issues that have been raised include the fact that not all workers who are eligible to make a claim will do so, and the fact that the likelihood of making a claim may vary across subgroups. Results of a survey conducted in one state (New South Wales) in 1993 indicate that over half (53%) of the workers who reported being injured at work in the previous 12 months did not make a claim (Australian Bureau of Statistics, 1994). Approximately 65% of these were either not covered by workers' compensations or the injury was considered by the worker to be too minor to make a claim. Of particular concern were the 8% who feared retrenchment, the 3% who were concerned about 'what others may think' and the 14% who were either not aware of workers' compensation or mistakenly thought they were not eligible.

These survey data on the likelihood of making a claim do not take into account the severity of the injury. It seems reasonable to assume that the likelihood of making a claim will increase as the severity of the injury, as indicated by the number of days off work, increases. Although the NDS is criticized for not including claims for less than five days' lost time, the data for these less severe cases will be more subject to under-reporting. In fact, it can be argued that imposing a threshold such as five or 10 days serves to focus attention on the more reliable and interpretable data. Another important consideration is the fact that compensation and rehabilitation policies are subject to change over time thereby making trends difficult to interpret.

Despite the limitations discussed above, compensation data are one of the few sources that identify work-related injuries, give some idea of the extent of the problem and provide information on some important associated factors. The development of a national data set for compensation-based statistics has been an important step in developing a national picture of occupational injury in Australia.

It should also be noted that, in addition to the state- and territory-based compensation schemes, some industries such as the coal industry maintain comprehensive accident/injury/incident reporting systems which provide data on all injuries, not just lost-time injuries (Joint Coal Board, 1995; Leigh *et al.*, 1990). Such data allow detailed analysis of occupational injury at the industry level.

1.6.2 Hospital in-patient data

Hospital in-patient records are another potential source of information. Under the ICD-9 classification system, the E-code for 'external causes' can be used to identify injuries. However, a fundamental deficiency of the current coding of hospital separations is that no item identifies a case as work-related and none provide any information on occupation or industry of employment. The likely introduction of an updated version of the ICD classification system (ICD-10) in 1998 in Australia may result in the adoption of an 'activity' classification which has a category for paid work and another for unpaid work. A 'place of occurrence' classification, used at present under ICD-9, has categories for 'industrial premises', 'mine or quarry' and 'farm'. While this is a guide to work-relatedness, it must be recognized that many work-related injuries occur in places such as roads and domestic premises. Moreover, for hospital separations data in Australia, the 'place of occurrence' item is missing in a large minority of cases. Adoption of the ICD-10 'activity' classification and improvements in the data collected for 'place of occurrence' could make these data a useful source, particularly for the farming sector where many workers are self-employed and hence are not covered by workers' compensation. Without these changes, hospital in-patient data are likely to remain a potentially useful source of information on work-related disorders that cannot be appropriately utilized for monitoring and prevention activities.

1.6.3 Hospital emergency department data

Hospital emergency departments deal with a wide range of injury cases. Unlike workers' compensation schemes, the hospital system can be accessed by all workers. The Injury Surveillance Information System (ISIS) was established in 1986. Although this system collects very detailed information on injury cases and has been used extensively for the study of factors associated with injury, only about 50 hospitals have participated (Harrison and Tyson, 1993). The data do not provide a representative national picture and have limited value for statistical reporting. Participation rates have declined since 1991 because of the high cost of obtaining the data and difficulty in collecting information. A particular limitation of ISIS has been the use of specially designed computer software that cannot be integrated with the computer-based information systems being introduced into hospitals. Nevertheless, approximately 700000 records have been collected, which provide a rich source of data for the study of injury causation.

The recent development of a National Minimum Data Standard for Injury Surveillance provides a basis for the collection of national injury data and is designed to be incorporated into information systems being introduced in hospitals (National Injury Surveillance Unit, 1995). Major considerations for designing the standard are that it should provide data of central importance to injury prevention practitioners, be small and simple to use in order to facilitate its incorporation into routine systems, be compatible with coding standards such as ICD-9 and be capable of providing reliable and valid data.

An important feature of this standard is that it includes three levels. The first is designed for data collection on all cases, the second is an expanded version of the first for collecting more detailed data on a sample of cases and the third comprises specialized items and classifications needed to conduct in-depth studies for certain types of injury. Level one will include items for main external cause of the injury, intent, type of place, type of activity, nature of injury, body region, occupation and a short narrative description. Socio-demographic information is also included. Although only a limited number of codes are available at level one, work-related cases can be identified and the inclusion of an item for intent will allow the identification of cases of violence in the workplace, which is regarded by some as an important emerging OHS issue. Although violence cases are represented in workers' compensation data, the coding does not allow them to be identified. Level two comprises a more detailed coding scheme for the items at level one, an extended narrative and also includes additional items concerning major factors associated with the injury, breakdown events and mechanism of injury.

The structure of this standard takes into account the fact that it is not feasible to collect detailed information on the large number of cases which pass through a hospital emergency department. The approach of collecting more detailed information on a representative sample of cases is a major strength. Although level one of the standard is being implemented at present in one state, it is not clear to what extent it will be adopted nationally and what the time-scale for this will be. In practice, achieving representative coverage at level two may be difficult to attain and obviously depends on the successful implementation of level one.

If successful, this surveillance system will provide a valuable source of information to complement workers' compensation data. The inclusion of a data item at level two to identify self-employed workers could also provide a means of directly comparing the experiences of these workers to wage and salary earners. In the long term these data could be used to provide a statistical basis for monitoring injury rates over time and also to study factors associated with work-related injuries.

1.7 DENOMINATORS

For traumatic fatalities, it is possible to obtain accurate data on the number of cases, sometimes referred to as the 'numerator'. However, interpretation of these data is often difficult in the absence of reliable data on the number of workers at risk (the denominator). We usually rely on census data or labour-force surveys to provide estimates for the number of workers in occupational, industry and socio-demographic groups. In practice, we tend to focus on the difficulties of obtaining accurate numerator data and lose sight of the fact that the denominator data, such as census data for example, are often out of date and also inadequate for some occupational groups such as seasonal workers and multiple job holders. Although labour-force surveys are conducted on a regular (ongoing) basis, denominator estimates based on these surveys are subject to random variability. Estimates for small subgroups of workers can be unreliable. In some situations data from various sources can be combined to develop the most appropriate denominators for a particular analysis (Corvalan et al., 1994; Driscoll et al., 1995). However, it must be recognized that, even though it may be possible to obtain accurate data for some outcomes, deficiencies in the available denominator data can introduce bias and random error when calculating rates.

1.8 OBTAINING AN OVERVIEW FROM MULTIPLE DATA SOURCES

Although each of the data sources described above provides valuable information, each covers only a subset of cases and hence can only describe occupational injury rates from one perspective. Even though they have some data items in common and there is compatibility in coding at a broad level, obtaining an overview by combining information across these sources is difficult. The inclusion criteria vary between these sources but there are nevertheless significant overlaps in coverage. For example, not all work-related cases present to a hospital emergency department, and not all work-related cases make a successful workers' compensation claim above the five-day threshold for the NDS. However, many cases could appear in both data sets.

In principle, linkage between data files provides a means of combining data from two or more sources and thereby extracting further valuable information. By linking records for an individual across data sets, the information for that person can be brought together in a single file. This would allow persons who are represented in multiple data sets to be identified and also provide more extensive information on such cases. Although this approach provides a mechanism for making the most effective use of the available data, in practice opportunities for linking files are severely limited by the lack of a unique personal identifier on data files and major community concerns in Australia regarding privacy (Sibthorpe *et al.*, 1995).

1.9 CONCLUSIONS

The importance of national data standards as a necessary requirement for the development of national information systems has been recognized in Australia and such standards have been developed in the OHS and health sectors. However, successful implementation of these standards is often difficult to achieve and also to maintain over time.

Traumatic fatalities represent a special case for which accurate and detailed information can be collected on almost all cases. Death is an extremely important and also clearly defined outcome. Compared with non-fatal injuries there are a relatively small number of cases and hence it is feasible to identify nearly all cases and collect extensive data on each. Even though the NDS only contains information on the more severe cases of work-related injury, it is not possible to collect information to the same level of detail on these 160 000 cases per year as it is to collect data on the 500 fatalities per year.

For non-fatal injury, none of the data sources discussed can provide an overview of work-related injury. Whilst the NDS standard seeks to collect the same level of information on all cases, the approach adopted in the standard for hospital emergency department data takes into account the trade-off between the number of cases and the depth of information collected by seeking more detailed information for only a sample. In principle, this approach can provide valid statistical data as well as a basis for studying causality. Although the framework that has been developed for injury surveillance through hospitals potentially provides a valuable source of information, it remains to be seen how well and how widely it will be implemented. Developing a data standard does not guarantee that it will be adopted universally. Also, standards are not static and are subject to change over time, which in some cases may compromise our ability to monitor trends.

Despite the methodological and statistical problems associated with obtaining a statistical overview of occupational injury, the available data do provide valuable information for surveillance purposes. The proposed development of new computerized data-collection systems promises to expand our current information base for work-related injuries and also to make such data more accessible for the purposes of surveillance, monitoring trends, developing standards and identifying priorities for prevention activities.

References

AUSTRALIAN BUREAU OF STATISTICS (1994) *Work-related injuries and illnesses: New South Wales, October 1993.* ABS Catalogue No. 6301.1.

CORVALAN, C.F., DRISCOLL, T.R. and HARRISON, J.E. (1994) Role of migrant factors in work-related fatalities in Australia. *Scandinavian Journal of Work, Environment and Health*, **20**, 364–370.

DRISCOLL, T.R., ANSARI, G., HARRISON, J.E., FROMMER, M.S. and RUCK, E.A. (1994) Traumatic work-related fatalities in commercial fishermen in Australia: *Occupational and Environmental Medicine*, **51**, 612–616.

(1995) Traumatic work-related fatalities in forestry and sawmill workers in Australia. *Journal of Safety Research*, **26**, 221–233.

ERLICH, S.E., DRISCOLL, T.R., HARRISON, J.E., FROMMER, M.S. and LEIGH, J. (1993) Work-related agricultural fatalities in Australia, 1982–1984. *Scandinavian Journal of Work, Environment and Health*, **19**, 162–167.

FEYER, A.M. and WILLIAMSON, A. (1991) The role of work practices in occupational accidents. *Proccedings of the Human Factors Society*, **35**, 1100–1104.

HARRISON, J. and TYSON, D. (1993) Injury surveillance in Australia. *Acta Paediatrica Japonica*, **35**, 171–178.

HARRISON, J.E., FROMMER, M.S., RUCK, E.A. and BLYTH, F.M. (1989) Deaths as a result of work-related injury in Australia, 1982–1984. *Medical Journal of Australia*, **150**, 118–125.

INDUSTRY COMMISSION (1995) *Work, health and safety: an inquiry into occupational health and safety*. Report No. 47, Canberra: AGPS.

JOINT COAL BOARD (1995) *Lost-time injuries and fatalities: New South Wales coal mines 1994–95*, Sydney: Joint Coal Board.

LEIGH, J., MULDER, H.B., FARNSWORTH, N.P. and MORGAN, G.G. (1990) Personal and environmental factors in coal mining accidents. *Journal of Occupational Accidents*, **13**, 233–250.

MACASKILL, P., MANDRYK, J. and LEIGH, J. (1995) *Improved coverage and relevance of national occupational health and safety surveillance in Australia*. Worksafe Australia National Institute Report, Canberra: AGPS.

MANDRYK, J.A., HARRISON, J.E., DRISCOLL, T.R. and RUCK, E.A. (1996) Forklift-related fatalities in Australia, 1982–84. *Journal of Occupational Health and Safety – Australia and New Zealand*, **12**, 219–227.

MOLLER, J. (1994) *Coronial information systems: needs and feasibility study*, Canberra: National Injury Surveillance Unit, Australian Institute of Health and Welfare.

NATIONAL HEALTH DATA COMMITTEE (1995) *The national health data dictionary*, version 4.0, Canberra: Australian Institute of Health and Welfare.

NATIONAL INJURY SURVEILLANCE UNIT (1995) *National data standards for injury surveillance*, version 2.0, Canberra: Australian Institute of Health and Welfare.

NATIONAL OCCUPATIONAL HEALTH AND SAFETY COMMISSION (1987) *National dataset for compensation-based statistics*, Canberra: AGPS.

SIBTHORPE, B., KLIEWER, E. and SMITH, L. (1995) Record linkage in Australian epidemiological research: health benefits, privacy safeguards and future potential. *Australian Journal of Public Health*, **19**, 250–256.

STOUT, N. and JENKINS, E. (1995) Use of narrative text fields in occupational injury data. In *International collaborative effort on injury statistics*, vol. 1, DHHS Publication No (PHS) 95-1252, Hyattsville, MD: United States Department of Health and Human Services.

WORKSAFE AUSTRALIA (1994) *Compendium of workers' compensation statistics 1991–92*. National Institute Report, Canberra: AGPS.

(1995) *Compendium of workers' compensation statistics 1992–93*. National Institute Report, Canberra: AGPS.

CHAPTER TWO

Analysis of narrative text fields in occupational injury data

NANCY STOUT

The purpose of this chapter is to demonstrate the uses and value of narrative text fields in occupational injury data and encourage the inclusion of narrative data in occupational injury surveillance systems.

At the National Institute for Occupational Safety and Health (NIOSH) we use surveillance data on occupational injuries and deaths to target high-risk worker populations and identify injury risk factors. Our analyses drive injury prevention efforts and impact legislation and regulatory policy. The inclusion of narrative fields in injury surveillance data allow us to identify specific hazards and injury cases that we would not otherwise find.

NIOSH has two national surveillance systems for occupational injuries. For fatal injuries we maintain the National Traumatic Occupational Fatalities (NTOF) surveillance system (Jenkins *et al.*, 1993), which comprises information from death certificates. All 50 states provide death certificates for cases that meet the criteria: 16 years of age or older, external cause of death (ICD9 E800-E999), and the 'injury at work' item marked 'yes'. One reason NTOF is so useful is that it contains narrative data that allow us to examine detail that is not typically available because most databases solely consist of coded data. NTOF contains narrative entries for industry, occupation, an injury description and immediate, underlying and contributory causes of death.

For non-fatal injuries we use the National Electronic Injury Surveillance System (NEISS), which comprises records from a nationally representative sample of 65 hospital emergency departments. Cases of injuries that occurred on the job are reported to NIOSH (Layne *et al.*, 1994). NEISS contains narrative fields for industry, name of business and occupation, and two injury description fields.

One simple example of how narrative data allow us to understand the circumstances of an injury better is this case from NEISS. The description from the coded data indicates that a 37-year-old male received a puncture in the lower leg, the source of injury was a nail and the injury event was that he was struck by a flying object. It appears that this person punctured his leg while hammering a nail. However, when we review the narrative injury description, we see a very different scenario: 'co-worker was using a nail gun and nail went through wall and struck patient in leg'. These circumstances, involving a co-worker and a nail gun, are quite different from what the coded data led us to believe and would require a different approach to prevention.

Analyses of these narrative entries, through computerized keyword searches and manual review, have allowed us to go beyond the limits of coded data to understand better specific circumstances and risks. Examples of analyses, many of which have impacted national programs or policies, are described below to demonstrate the value of these narrative data.

First, several deaths were brought to our attention of farmers who had entered their manure pits and been overcome and died from methane gas. Tragically, other family members and co-workers were also dying in rescue attempts. We realized that we needed to alert the public to the risk of death due to methane asphyxia in manure pits on farms. Although we had investigated a few cases, we wanted to try to determine the national magnitude of this problem. We therefore subset potential cases from the NTOF data by assigning to them appropriate ICD-9 E-codes (World Health Organization, 1977), such as suffocation and poisoning, which narrowed us to 2000 cases. We then manually reviewed the injury descriptions to confirm cases that occurred in manure pits.

Although the death certificates did not always contain enough detail to identify all cases, we found that at least 16 people died from asphyxia in manure pits between 1980 and 1985, and that five episodes resulted in multiple deaths from rescue attempts. We published an article on this and issued an Alert, which was disseminated to farmers across the country through county agricultural extension agents (Myers and Casini, 1989; US Department of Health and Human Services, 1990).

Another example resulted from a request from the Occupational Safety and Health Administration (OSHA), for information on hazards in the logging industry. While NIOSH is a research agency, OSHA is a regulatory agency that develops and enforces occupational safety and health regulations. NIOSH frequently provides data to OSHA when they are proposing new rulemaking for occupational safety. This is important because this is one way our findings can be used to shape national policy to prevent occupational injuries. We selected cases coded as 'logging industry' and examined the distribution of cause of death. The leading causes were deaths from falling objects and machinery. We then subset cases coded to the falling object and machinery E-codes, and reviewed the injury descriptions to understand better the circumstances of these deaths.

We found that 41% of the machinery-related deaths in logging were results of roll-overs, followed by run-overs (16%), being caught in the machinery (10%) and cable breaks (8%). Nine-tenths of the deaths due to being struck by falling objects were due to falling trees and logs. While that may not be surprising, we also found that of all worker deaths E-coded as being struck by falling objects, in all industries, 30% were from trees alone. Again, we could not have learned any of this from coded data. Without the narrative data, we would not have a clue as to what falling objects were killing workers or how we could attempt to prevent these deaths.

In addition to providing these findings to OSHA to support their proposed safety standards for logging, we combined these findings with results of field investigations of logging deaths to produce an Alert, which describes the risks that loggers face and provides recommendations for prevention (US Department of Health and Human Services, 1994a).

In response to many requests for information on forklift deaths, we conducted another analysis of narrative data from NTOF. Cases coded as cause of death due to machinery and containing keywords 'forklift', 'lift truck', etc. were selected through a computer search. We automatically classified cases into categories of types of incidents, using keywords such as 'roll-over', 'run-over', 'tip over', etc., then classified the remaining cases through manual review. We found that at least 563 workers died in forklift incidents from 1980 to 1988, and the vast majority were due to tip-overs. These data are being used by OSHA to support their proposed rule for Power Industrial Truck Operator Training.

Another example resulted from a congressional request for data illustrating the major occupational safety hazards related to agricultural machinery. We first selected cases with E-codes for agricultural machinery, then used keywords to search and subset the cases for the various common types of agricultural machinery (e.g. 'tractor', 'auger', 'baler'). Finally, we manually reviewed the injury descriptions to classify the remaining cases. Tractors accounted for 69% of all agricultural machinery-related deaths from 1980 to 1986 (1523 cases); 52% of these resulted from roll-overs and 16% were run-overs. Later, in an analysis of narrative data on motor vehicles, we discovered that many tractor incidents are missed by limiting analysis to the E-code for agricultural machines because, according to ICD-9 rules, tractors are correctly coded as motor vehicles if they are on a public roadway. We therefore undertook a new analysis to more comprehensively identify tractor deaths. Without limiting analysis to the E-code for agricultural machinery, we subset all the cases in the agricultural industry to the keyword 'tractor', and further subset these cases using the keywords 'roll-over' and 'run-over'. We discovered that 15% of tractor incidents had previously been missed by limiting analysis to the E-code for agricultural machinery. Ten per cent of the tractor deaths were (correctly) E-coded as motor vehicle deaths and 1% (14 cases) as being struck by a falling object (Jenkins and Hard, 1992). Tractor deaths are correctly coded as being struck by a falling object or other machinery if they are stationary, such as during maintenance and repair.

We also used narrative data to identify and understand worker deaths caused by falls from suspension scaffolds (US Department of Health and Human Services, 1992), from trench cave-ins (US Department of Health and Human Services, 1993), from falls through skylights and roof openings (US Department of Health and Human Services, 1989), from electrocutions during work with scaffolds near overhead power lines (US Department of Health and Human Services, 1991) and from entanglements with hay bailers that resulted in scalpings (Roeig et al., 1992; US Department of Health and Human Services, 1994b), and to determine the risks of injuries to adolescent workers (US Department of Health and Human Services, 1995). All of these analyses resulted in the publication and dissemination of information on the hazards associated with these circumstances and recommendations for the prevention of similar deaths.

These are just a few examples of the uses and value of narrative data. None of these analyses would have been possible with coded data, as none of these cases are identifiable through E-codes or other coded data. The narrative information, particularly the injury description, allows us to go beyond the limits of coded data to drive research and policy to prevent worker deaths.

The value of E-codes in addition to narrative data must also be emphasized. In many of these examples the E-codes allowed us to subset a large database to a manageable number of cases for manual review or to determine which words to use in keyword searches. Without this ability, many cases may be missed and statistical analysis of subsets would be not be possible. Coded and narrative data are complementary in injury surveillance data and, used together, allow us to learn much more about specific hazards and risks than either alone.

Another value of narrative data is the ability to code or recode variables to alternative coding schemes. For example, there are numerous coding schemes for occupation, industry and injury circumstances. Most schemes are not comparable and data coded to one cannot be directly converted to another. This prohibits comparisons between databases, particularly international comparisons. It also often prevents the computation of rates if denominator and numerator databases are coded to different schemes (which is frequently the case in the US). Coding schemes also change periodically. US employment codes generally change every decade and even internationally standardized codes such as the ICD are periodically modified. When surveillance systems convert to a new code structure, the ability to monitor trends over time is often lost.

Narrative data can be recoded to provide comparability of data between systems, years and countries. There have been a number of efforts in recent years to develop software that automatically codes narrative data into numeric codes. For example, NIOSH, in collaboration with several other agencies and organizations, has been developing a software system that automatically codes occupation and industry narratives into standard numeric codes. The Standardized Occupation and Industry Coding (SOIC) software system is a PC, windows-based coding system that uses a combination of relational database and artificial intelligence techniques to automatically code narrative descriptions of industry and occupation into a standard numeric coding scheme (US Department of Commerce, Bureau of the Census, 1982). A brief description of the coding engine is illustrative of this other valuable application of narrative data (Figure 2.1).

The SOIC coding engine is composed of a series of increasingly complex processes ranging from a simple look-up table to a comparison with all of the phrases in the *Alphabetic Index of Industries and Occupations* (US Department of Commerce, Bureau of the Census, 1982). This structure allows the system to be very flexible and efficient in its approach to coding. Each process attempts to code the literal independently and subsequent processes are used only if a code assignment cannot be made.

First, an autospell correction process corrects common misspellings and expands acronyms and fused words. The first coding process uses a table of *paired phrases* for quick, efficient look-ups of the most commonly occurring industry – occupation combinations. The primary reason for including this process is speed; the commonly occurring cases are quickly coded without having to perform detailed syntactical analyses.

The second process uses *idiom matching* data tables in order to identify specific occurrences of phrases that are, by their nature, very difficult to code without additional context-specific information. This process codes error-prone, hard-to-code phrases, such as 'aviation construction', a manufacturing industry. Evaluated separately, these words would be likely to lead to different industry codes.

Much like idiom matching, *company matching* is primarily a table of company names that do not have other specific information identifying the industry (e.g. Tarus Exploration – a mining company). This process also incorporates state-specific company names.

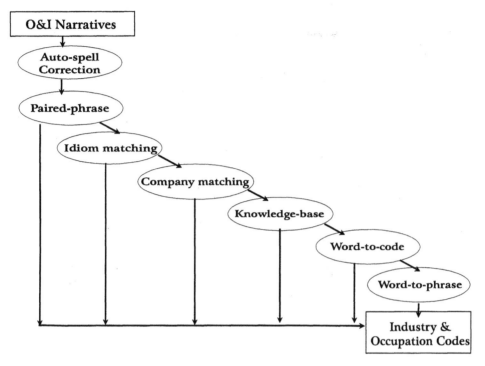

Figure 2.1 Standardized occupation and industry coding (SOIC) system.

The next process is *knowledge-based* coding. This is the heart of the coding engine and the artificial intelligence application. It consists of an extensive rule base that attempts to categorize literals into industry types and keeps narrowing these industry types into more specific industry groupings until a code assignment can be made. This process follows the same complex set of rules that expert coders follow to code difficult cases. It is one of the most flexible coding processes and can easily be expanded to deal with a wider range of phrases and codes.

The *word-to-code* process is based on the probabilistic relationship between a word and a given code (i.e. the probability that a given word will lead to a specific code), derived from analyzing data that were manually coded by an expert. It is much more flexible than some of the previous processes because it can handle different forms of an input literal, such as phrases with variable word order.

Finally, *word-to-phrase* coding analyzes the phrases in the *Alphabetic Index* to determine which phrase most closely matches the industry or occupation literal. This coding method is last because it performs an exhaustive, computation-intensive search of all potential matches in the *Alphabetic Index*.

Although this is a very brief overview of a complex process, it illustrates another valuable use of narrative fields. Automated coding software makes it easier, less expensive and more reliable and consistent to automate narrative data than to code variables prior to automation. With the state of current technology, automated coding systems are becoming more efficient and available. Moreover, as new coding software is developed, historical data that contain narratives can be coded to enable analyses that were not previously possible.

In the past, computer costs and memory limitations often prohibited the inclusion of long character fields in databases. Today, these limitations have been overcome and costs are minimal, particularly in comparison with the cost of expert coding. Narrative data not only provide valuable detail but also provide the flexibility to adapt existing data to changes and future needs. For all of these reasons, the inclusion of narrative text fields in surveillance systems and databases is infinitely valuable and strongly encouraged.

References

JENKINS, E.L. and HARD D.L. (1992) Implications for the use of E-codes of the International Classification of Diseases and narrative data in identifying tractor-related deaths in agriculture, United States, 1980–1986. *Scandinavian Journal of Work, Environment and Health*, **92**(18), suppl 2, 49–50.

JENKINS, E.L., KISNER, S.M., and FOSBROKE, D.E., *et al.* (1993) *Fatal injuries to workers in the United States, 1980–1989: a decade of surveillance, national profile*. DHHS Publication (NIOSH) 93-108. Cincinnati, OH: National Institute for Occupational Safety and Health.

LAYNE, L.A., CASTILLO, D.N., STOUT, N. and CUTLIP, P. (1994) Adolescent occupational injuries requiring hospital emergency department treatment: a nationally representative sample. *American Journal of Public Health*, **84**(4), 657–660.

MYERS, J.R. and CASINI, V.J. (1989) Fatalities attributed to methane asphyxia in manure pits – Ohio, Michigan, 1989. *Centers for Disease Control. Mortality and Morbidity Weekly*, **38**(33), 583–586.

ROEIG, S., MELIUS, J., CASINI, V.J., MYERS, J.R. and SNYDER, K.A. (1992) Scalping incidents involving hale balers – New York. *Centers for Disease Control. Mortality and Morbidity Weekly*, **41**(27), 489–491.

US DEPARTMENT OF COMMERCE, BUREAU OF THE CENSUS (1982) *1980 census of population: alphabetic index of industries and occupations*. Publication PHC80-R3.

US DEPARTMENT OF HEALTH AND HUMAN SERVICES (1989) *NIOSH alert: preventing worker deaths and injuries from falls through skylights and roof openings*, DHHS (NIOSH) Publication No. 90-100.

(1990) *NIOSH alert: request for assistance in preventing deaths of farm workers in manure pits*, DHHS (NIOSH) Publication 90-103.

(1991) *NIOSH alert: request for assistance in preventing electrocutions during work with scaffolds near overhead power lines*, DHHS (NIOSH) Publication No. 91-110.

(1992) *NIOSH alert: request for assistance in preventing worker injuries and deaths caused by falls from suspension scaffolds*, DHHS (NIOSH) Publication No. 92-108.

(1993) *NIOSH update: NIOSH warns of danger of trench cave-ins*, DHHS (NIOSH) Publication No. 93-110.

(1994a) *NIOSH alert: request for assistance in preventing injuries and deaths of loggers*, DHHS (NIOSH) Publication No. 95-101.

(1994b) *NIOSH alert: request for assistance in preventing scalping and other severe injuries from farm machinery*, DHHS (NIOSH) Publication No. 94-105.

(1995) *NIOSH alert: request for assistance in preventing deaths and injuries of adolescent workers*, DHHS (NIOSH) Publication No. 95-125.

WORLD HEALTH ORGANIZATION (1977) *International classification of diseases: manual on the international statistical classification of diseases, injuries, and causes of death*, 9th revision, Geneva, Switzerland.

CHAPTER THREE

Using injury data to identify industry research priorities

LAURIE STILLER, JOHN SARGAISON AND THOMAS MITCHELL

3.1 INTRODUCTION

Occupational health and safety (OHS) research plays an essential role in reducing the exposure of Australian coal-mine workers to occupational risks. Presently there are considerable financial resources allocated to OHS research in coal-mining, however, the Australian Coal Association (ACA) has commented that in the past 'because so many organizations are involved in relevant research it has been difficult to produce a coordinated strategy which will produce the best results with the minimum overlap and duplication' (Australian Coal Association, 1992).

To address this issue, the ACA, in conjunction with BHP Australia Coal, commissioned a study of OHS in the Australian Black Coal Industry. The Victorian Institute for Occupational Safety and Health (VIOSH) Australia was funded to review available injury and disease data by the Australian Coal Association Research Program (ACARP) and the BHP Australia Coal Special Research Program in collaboration with the Joint Coal Board (JCB) Health and Safety Trust. The project was also designed to include an analysis of the status and nature of local and international coal-specific occupational health and safety research.

The general strategy involved analysis of injury and disease data from the New South Wales (NSW) and Queensland coal industry for the period 1 January 1991 to 1 January 1993 using a severity index (reflecting compensation costs) to identify general OHS problems for a range of occupational groups including underground miners, supervisors, maintenance workers, opencut miners and all other surface mine workers. From the general research recommendations, it was necessary to identify *specific* research projects capable of delivering practical solutions. To help further focus the research strategy, a research vision 'to eliminate all deaths and cases of permanent disability in the Australian coal industry' was adopted. The data for injuries of 30 days or longer in duration and fatality data were then re-analysed to identify those items of equipment most commonly linked with severe outcomes.

This chapter gives an overview of the methodology and some illustrations of how injury data were used to achieve this aim.

3.2 PHASE 1: VIOSH RESEARCH

A full description of this part of the project is reported elsewhere (Mitchell and Larsson, 1994). The following provides a brief overview.

3.2.1 Analysis methodology

Assorted accident and workers' compensation data relating to injuries and disease in the Australian Black Coal Industry were analysed. The main body of data was drawn from the claims experience period from January 1991 to June 1993. The analysis was performed by ranking the risk associated with injury and illness classifications used by the industry. The ranking process considered:

- the prevalence of injury and ill-health categories;
- the severity of the injury and disease distinguished by classification; and
- the number of Black Coal Industry workers exposed to the problem.

In preparing their analysis, Mitchell and Larsson (1994) suggested that analysis of workers' compensation claims data for prevention purposes is a process of stepwise selection. This selection process is based on the assumption that it is more important to identify exposures associated with medically and economically severe injuries than it is to identify the factors contributing to relatively minor medical and economical losses. The optimal prevention strategy, however, will always have to find a compromise between rather uncommon severe trauma, with high degrees of medical severity and substantial losses to the individual, and large amounts of less severe ailments with limited consequences to those injured and with major expense to the organization.

There are a number of different ways of defining and measuring costs in relation to occupational injury (Andreoni, 1986). To set priorities for prevention, the consequences of occupational trauma, as defined and compensated by occupational injury insurance, has proven very useful (Larsson, 1990, 1991). There are also examples of studies from the mining industry that have analysed the costs related to occupational injuries to identify prevention strategies (Chi and Hamilton, 1986; Bhattacherjee et al., 1992).

3.2.2 Treatment of compensation information

A data set of 16 277 claims were identified by the NSW JCB as being registered with an occurrence date between 1 January 1991 and 30 June 1993. Scrutiny of these claims by the NSW JCB reduced the data set to 14 432 records after registered claims with no compensation payment (or payments totalling zero) and claims with payments made after 30 June 1993 were removed.

3.2.3 Reconstruction of compensation data set

The data were sorted into three claim types:

- injuries with no lost time;
- injuries with lost time; and
- fatalities.

Analysis strategy varied with the type of incident being considered. Data from the first and the third categories, injuries with no lost time and fatalities, were analysed in relation to the total industry experience for the study period. Data from the second category, injuries with lost time, were categorized under broad occupational groups. For the purpose.of this study six broad occupational groups were constructed for examining injuries with lost time data:

- underground face workers;
- underground supervisors;
- surface excavation workers;
- trades and allied workers;
- washery workers; and
- administrative staff, management and surface support workers.

The analysis of the injuries with lost time data described occupational groups in relation to three general factors:

- age of claimants;
- compensated amount per claim; and
- duration of time lost.

The injuries with lost time data for the six occupational groups were further analysed in relation to eight descriptive incident variables to provide an insight into the major injuries/disease experienced by the workers of each group. The descriptors with their corresponding NSW JCB database variable field heading were:

- occupation (occupation code);
- injury (injury nature code);
- part of body (body code);
- task (task code);
- place (place code);
- agency (injury agency code);
- mechanism (incident type code); and
- grouped variable.

3.2.4 The severity index

To identify specific tasks, equipment or injuries associated with more severe consequences (and thus identify priorities for intervention), an index measure of relative severity was constructed from the various types of compensation paid for each claim. This severity index was used to analyse the variables listed above. The severity index contains four components referred to as compensation, lump sum, medical and other.

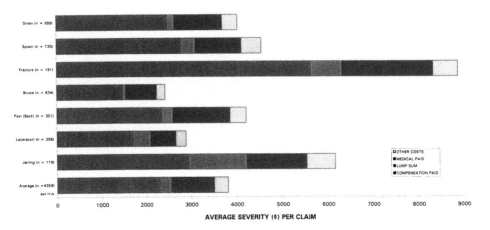

Figure 3.1 Typical analysis of database variables using the severity index.

Compensation represents the amount of compensation paid (either in full and/or partial amounts) for loss of earnings. Because the compensation component relates to time lost by the worker, it is therefore able to mirror directly the length of the rehabilitation period. Lump sum indicates the amount paid (or held in reserve) in relation to medical impairment and ensuing permanent disability according to the table of maims or as the result of litigation. The third part of the index, medical, contains the sum of the amounts paid for hospital, medical treatment and related rehabilitation expenses. In combination, lump sum and medical amounts reflect the medical severity of variable being considered. The final component, other, represents all other forms of miscellaneous compensation categorized by the scheme. Figure 3.1 shows a sample of the analysis undertaken using the severity index to identify specific risk situations.

In a separate analysis using the severity index, 848 injuries with a lost time of more than 30 days were extracted from the data. As described earlier, these incidents were classified into occupational categories and in terms of descriptors of the incident. The purpose of this analysis was to identify general OHS problems resulting in substantial disability for the various occupational groups. The general areas identified included musculo-skeletal injuries, mine-environment-related injuries, equipment operation injuries and noise-induced hearing loss. Many of these areas were identified as being inadequately addressed in either local or international current research. Thus, although Phase 1 resulted in a list of priority OHS issues for each occupation group, further analysis was needed to distil specific research (projects) from the data.

3.3 PHASE 2: INDUSTRY RESEARCH

3.3.1 Analysis methodology

Following the completion of the Mitchell and Larrson (1994) study (Phase 1), the Coal Industry Research Strategy objectives were further refined to focus on 'eliminating death and permanent disability'. This focus was influenced by McDonald's (1996) work on a three-tier injury classification system in which:

Table 3.1 Equipment involved in most severe injuries

Underground equipment	Open cut equipment
Loaders	Dozers, tracked
Continuous miners	Loaders
Man transporters	Light vehicles
Conveyors	Draglines, buckets
Roofbolters	Scrapers

■ Class I damage permanently alters a person's future (fatality or permanent impairment);

■ Class II damage temporarily damages a person's future (temporary partial impairment); and

■ Class III damage inconveniences a person's future (minor partial impairment).

McDonald (1996) stresses that OHS is a Class I problem because Class I events are those resulting in the most human suffering and the highest costs.

Based on the Phase 1 research data, *underground* and *open-cut* equipment associated with the most severe injuries were identified as shown in Table 3.1. Data for each of these equipment items over the period 1 January 1991 to 30 June 1994 were further analysed to identify causal factors better. This analysis addressed injuries of 30 hours' (approximately 5 days) duration with a particular emphasis on the more serious injuries of 240 hours' (approximately 30 days) duration. Again, data concerning injuries requiring 30 days lost were selected because, in the absence of better data, it was assumed that these injuries are the ones most likely to reflect 'permanent disability' incidents.

Descriptive information for each incident was used to develop unique classification systems for each item of equipment. Incidents of greater than 30 hours' and greater than 240 hours' lost time were categorized using the system. In this way, a taxonomy of all severe injuries that occurred between January 1991 and July 1994, relative to each of the identified equipment types, was developed to build a picture of exactly where the most serious incidents were occurring. Comparisons were also drawn with the number of short-term injuries (one week's duration) in each category. This enabled issues as specific as 'access and egress to loaders' and 'manual handling of continuous miner cables' to be pinpointed so that projects could be framed to address areas of concern amenable to research.

3.3.2 Major findings

The analysis revealed important issues concerning overall patterns of injury occurrence. Figure 3.1 shows the pattern of injury occurrence resulting in higher and lower levels of lost time in continuous miner incidents in NSW. A significant feature that emerged from these data was that although a downward trend was apparent in lost-time injuries of greater than 30 hours' duration (from 1991 to 1994) for each item of equipment, the same trend was not always evident in the more serious injuries of 240 hours' or more duration (Figure 3.2). This indicates that while the overall number of injuries is decreasing, the potential for serious injury (e.g. permanent disablement) is not decreasing. A focus for the research strategy on permanent

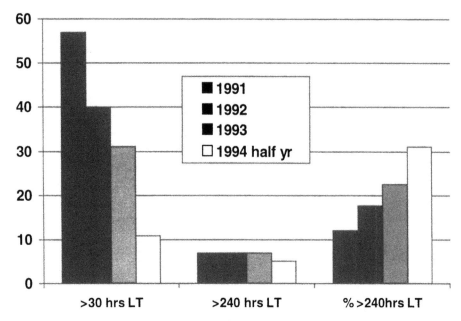

Figure 3.2 Continuous miner incidents in NSW.

disabling injuries and fatalities is therefore highly appropriate. It should be noted that these data represent gross numbers of injuries and are not adjusted for factors such as decreases or increases in the use of items of equipment.

There are perhaps several reasons why a downward trend is not apparent among more serious incidents. First, current prevention strategies may be rather reactive. This may result in the focus of preventive efforts being on high frequency, low severity events. Such a focus may not necessarily result in the causes of low frequency, high severity events being addressed. Second, improved rehabilitation and return-to-work practices are more likely have an impact on the less severe end of the injury spectrum. This may lead to an uneven distribution of improved injury outcomes.

The taxonomies of information describing all severe injuries were also found to be useful in identifying very specific aspects of equipment operation that could be related to high severity incidents. By compiling industry data in this way it is possible to pinpoint areas for attention that would not be identified from an analysis of an individual mine's data. A sample taxonomy for conveyor belts is shown in Table 3.2 and the summary of conveyor belt incidents in two states is shown in Table 3.3. These data highlight that inspection/cleaning, access/egress and manual handling in belt moves are the major areas of concern.

3.4 CONCLUSION

Compensation data were found to be a useful tool in identifying the prevalence and severity of traumatic occupational injuries suffered by coal-miners. The methods applied in Phase 1 were able to use the compensation data to identify broad risk

Table 3.2 Taxonomy of conveyor belt incidents

Main category	Sub-category	<240 hours', >30 hours' lost time		>240 hours' lost time	
		NSW	QLD	NSW	QLD
Maintenance	1.1 Manual handling	17	4	3	2
(changing rollers)	1.2 Struck by/caught in	6	1		
	1.3 Bump into	1		1	
	1.4 Slip/trip	4	1	1	
	1.5 Unplanned move	1	1		1
	1.6 Falling object	5			
Access/egress	2.1 Slip stepping off	7	3		2
	2.2 Slip climbing over	14	1	4	2
	2.3 Struck by/caught in				
	2.4 Slip/trip working on	7		3	1
	2.5 Hit by going under	2			
	2.6 Unplanned move	1		1	
	2.7 Bump into				
Operation	3.1 Struck by/caught in	5		3	1
(inspection/	3.2 Flying/falling object	3		1	
cleaning)	3.3 Slip/trip	2	2		2
	3.4 Unplanned move				
	3.5 Manual handling	2	1		1
Extend/retract	4.1 Manual handling	25	9	4	3
belt move	4.2 Struck by/caught in	5	5	2	3
	4.3 Bump into	2	1		
	4.4 Slip/trip		1		
	4.5 Unplanned move	1			1
	4.6 Falling object	2		1	

NSW = New South Wales; QLD = Queensland.

Table 3.3 Summary of results of taxonomy for conveyor belt incidents in NSW and Queensland

Main category	>30 hours lost-time injuries	>240 hours lost-time injuries	% >240 hours
Maintenance	54	8	14
Access/egress	56	13	23
Operation	32	8	25
Move	72	14	19
Total	214	43	20

areas within several occupational groups. The severity index was shown to be useful for this purpose as it emphasizes the costs associated with key claims categories.

However, there are inadequacies in the existing data. In particular, incident classification systems are developed to suit the workers' compensation systems rather than prevention and the needs of mines. From the present analysis it is clear that if specific projects capable of delivering practical solutions are to emerge from injury data, much more specificity is needed in the information to be considered.

The information needs to be specific both with respect to the characteristics of the injured and the characteristics of the factors involved.

Another major inadequacy of standard collection systems highlighted in this study was that incidents that resulted in Class I injuries and those that had Class I injury potential could not be accurately identified. In the present study a surrogate grouping was used, namely those injuries with longer duration lost-time outcomes. It is clear from the analysis of these data that one cannot directly and automatically extrapolate from lesser outcomes to more serious outcomes. The present findings suggest that reliable, accurate data on permanent disabling injury is critical to effective priority development.

Other enhancements to the system would also increase the accuracy and usefulness of the information. These are the provision of exposure information for all occupational categories represented in the industry to increase the ability to assess risk exposure and the use of the medical surveillance system to provide an up-to-date assessment of the prevalence and severity of occupational diseases.

In summary, the present project has demonstrated that injury data can be used effectively to guide the identification of research priorities in the coal industry. The form and content of the data sets are, however, critical. Industry data are most commonly used to attempt to benchmark safety performance with other companies in the industry or with other industries on the basis of lost-time frequency rates. This can be an unproductive and often counterproductive exercise. Not nearly enough companies use industry data (often because it is badly presented, is difficult to access and does not adequately focus on high severity incidents) to identify areas of high risk potential in their operations.

National injury data sets have been developed on the basis of what data is practically gatherable from workers' compensation systems rather than from the perspective of what data is needed to support the preventive efforts of companies. Taking this focus would generate a totally different data system from the one which is currently in use in most jurisdictions in Australia. A better data system has the potential to be a major driver of the prevention of death and permanent disabling injury, and this surely is our number one priority at state and national level.

References

ADREONI, G. (1986) *The cost of occupational accidents and diseases*, Geneva: ILO.

AUSTRALIAN COAL ASSOCIATION (1992) *Australian Coal Association Research Program Strategy*, Sydney: ACA.

BHATTACHERJEE, A., RAMANI, R. and NATARAJAN, R. (1992) Injury experience analysis for risk assessment and safety evaluation. *Mining Engineer*, December.

CHI, D. and HAMILTON, B. (1986) Trends in mining accidents and their costs (1975–1984). *Proceedings of the 19th International Symposium on the application of computer methods in the minerals industry*, Pennsylvania State University.

LARSSON, T. (1990) *IPSO Factum 21: accident information and priorities for accident prevention*, Stockholm: IPSO International.

(1991) *A review of the injury prevention potential of the ACC*, Wellington: IPSO New Zealand.

McDONALD, G.L. (1996) Focus don't fiddle (the obscenity of the LTIFR), Paper presented at the Occupational Injury Symposium, Sydney, February.

MITCHELL, T. and LARSSON, T. (1994) *Report of commissioned study of occupational health and safety in the Australian coal industry*, University of Ballarat Press.

PART TWO

Errors, mistakes and behaviour

INTRODUCTION

The involvement of the human element in accident and injury causation is ubiquitous. Most estimates of the actual involvement are very high, e.g. up to 80% of aircraft accidents are reported to involve human error (Schuckburgh, 1975) and 90% of fatalities in Australia in a three-year period involved human behaviour (Williamson and Feyer, 1990). However, the human element is a very broad category and simply counting its presence or absence is not very helpful in identifying its role or how one might intervene to modify its role in the occurrence of injury. An essential component of accident prevention therefore is gaining a better understanding of the nature, causes and consequences of errors and behaviour (Feyer and Williamson, 1997).

One of the important characteristics of behaviour is its adaptive nature. Errors are a normal part of behaviour and play a fundamental role in learning new skills and behaviours and in maintaining those behaviours (Reason, 1990). Through testing the boundaries of our interactions with the environment and consequently making errors we learn just what those boundaries are. This is essential for learning a new skill and also for updating and maintaining skills already learned. Arguably, the occurrence of injury involving a behavioural cause is the prima facie case of interaction with the environment and testing of the boundaries. Thus accidents themselves can be part of the shaping of subsequent behaviour. The chapter by Laughery and Vaubel takes up the issue of the relationship between accident history and subsequent accidents, and the adaptive changes that might occur as a consequence of injury. It reports that characteristics of injury (e.g. severity) and characteristics of the accidents (e.g. similarity of the tasks being performed when a first and a second accident are compared) are related to subsequent occurrence of accidents. The clear implication drawn by the authors is that behaviour is modified by prior accident experience, much in the way that would be expected as part of the learning process.

The critical role played by errors in learning and updating skills means that they are a constant feature of behaviour. Human error has often been used to encompass rather broadly all the ways in which the human factor can contribute to accidents at the time of the accident (Parker *et al.*, 1995). However, one of the most important aspects of the nature of error is that it is not a unitary phenomenon. Errors differ depending on the information processing functions being challenged, so that the error can signify breakdown in different information processing functions. Different types of errors can also be distinguished between skilled and unskilled behaviour. Errors during skilled behaviour occur as absent-minded or unintended actions and are quite distinct from the mistakes which occur during unskilled behaviour (Rasmussen, 1981; Reason, 1990; Norman, 1981). Apart from lapses and mistakes, unsafe actions can also be classified as violations where they are deliberate deviations from safe operating procedures (Reason, 1990). The distinctions between these different classes of human behaviour are far from hard and fast. However, such differences in the nature of the error have critical implications for interventions to prevent them.

The chapter by Pollock takes up the notion that knowledge of error forms has the potential to provide important insights about their causes and consequently appropriate remedial action. Moreover, she argues that information obtained about error occurrence in work settings may be a very sensitive early indicator of the adverse effects on performance, prior to the occurrence of accidents and injury. In particular, the approach is applied to evaluation of the impact of shiftwork on human performance. The effect of night work on errors and behaviour is well documented. Studies have shown consistently that errors are more common in the early hours of the morning, that is during night shift (Folkard and Monk, 1979). The risk of work-related fatalities occurring at night, corrected for the estimated numbers of workers at work, has also been shown to be considerably increased and more likely to be the result of behavioural factors (Williamson and Feyer, 1996). Not all adverse aspects of shiftwork, however, result in increased accident and injury rates. Pollock argues that other aspects of shiftwork, such as shift duration, may have subtle but highly specific effects on information processing and performance in the absence of un-changed accident and injury rates. The chapter concludes that examination of the nature of errors that occur may be used as a highly sensitive diagnostic marker of the adverse effects of aspects of work organization, particularly when work organization is changed.

The chapter by Hancock and Manser also considers the utility of understanding fundamental mechanisms in information processing if the demands inherent in work tasks are to be better understood. In particular, the authors report on theoretical issues and research findings associated with time-to-contact and the perceptual phenomena underlying this aspect of performance. The importance of such aspects of human performance in the causes of injury are quite evident. It is very likely, for instance, that the perceptual phenomena discussed in this chapter play an important role in a common mechanism of injury i.e. collision with moving objects. The most recent national statistics available in Australia indicate that contact with moving objects is the third most common mechanism of compensated injury or disease, accounting for approximately 13% of cases (National Occupational Health and Safety Commission, 1996). This chapter highlights that understanding of the basic parameters of human operator performance is essential for more effectively preventing errors and their potential consequences.

Human behaviour and error are complex phenomena about which we know a great deal. Systematic application of this knowledge to the role of human behaviour in occupational injury, however, is relatively recent. The chapters in this part focus on how such knowledge might be applied to understanding the determinants and outcomes of human behaviour in occupational injury.

References

FEYER, A.-M. and WILLIAMSON, A.M. (1997) Human factors in accident modelling. In *ILO encyclopaedia of occupational health and safety*, 4th edition, ed. J. Stellman. Geneva: ILO.

FOLKARD, S. and MONK, T.H. (1979) Shiftwork and performance. *Human Factors*, **21**, 483–492.

NATIONAL OCCUPATIONAL HEALTH AND SAFETY COMMISSION (1996) *Compendium of Workers' Compensation Statistics, Australia, 1994–95*, Canberra: Australian Government Printing Service.

NORMAN, D.A. (1981) Categorisation of action slips. *Psychological Review*, **88**, 1–15.

PARKER, D., REASON, J.T., MANSTEAD, A.S.R. and STRADLING, S.G. (1995) Driving errors, driving violations and accident involvement. *Ergonomics*, **38**(5), 1036–1048.

REASON, J.T. (1990) *Human error*. Cambridge: Cambridge University Press.

RASMUSSEN, J. (1981) Models of mental strategies in plant diagnosis. In *Human detection and diagnosis of system failures*, ed. J. Rasmussen and W. Rouse, New York: Plenum.

SCHUCKBURGH, J.S. (1975) Accident statistics and human factors. *Aviation, Space and Environmental Medicine*, **46**, 76–79.

WILLIAMSON, A.M. and FEYER, A.-M. (1990) Behavioural epidemiology as a tool for accident research. *Journal of Occupational Accidents*, **12**, 207–222.

(1996) Causes of accidents and time of day. *Work and Stress*, **9**(2/3), 158–164.

CHAPTER FOUR

The role of accident experiences on subsequent accident events

KENNETH R. LAUGHERY AND KENT P. VAUBEL

4.1 INTRODUCTION

The basic issues addressed in the research reported here concern how the experience of having an accident influences a person's having subsequent accidents. More specifically, it addresses the extent to which characteristics of accident events involving a particular employee in an industrial setting influence the occurrence of subsequent accidents for that employee. Certainly it seems reasonable to assume that having an accident will affect a worker in the sense that employees learn and modify their behaviour in some adaptive fashion.

Early studies exploring the relationship between accidents and subsequent behaviour focused on individual differences (e.g. Greenwood and Woods, 1919; Mintz, 1954; Mintz, 1956). Although prevalent, this approach was virtually abandoned due to its limited value in understanding the etiology of accidents. An example was the accident-proneness concept and the unsuccessful attempts made to identify 'accident-prone' workers. Some of the work (e.g. Mintz, 1954) employing the individual differences paradigm was methodologically important, however, because it employed time intervals between accidents as a dependent measure. Relative to simple frequency counts, use of intervals between accidents offers a useful way of capturing the effects of accidents on behaviour.

One can in part view the influence of occupational accidents on subsequent events as a learning phenomenon. In a work context, accidents can to some extent be thought of as the result of errors, and errors are closely related to the learning process. Indeed, for Rasmussen (1985), learning and error-making are closely related; people learn by detecting and correcting errors. From this perspective, job-related injuries serve as feedback about how well the individual is performing the job.

The use of feedback as a means to improve safety is not new. Longitudinal studies have demonstrated the positive influence of information about unsafe acts committed in the workplace on subsequent behaviour (Komaki et al., 1980; Komaki et al., 1982). Assessing the impact of feedback in the form of accidents that have actually occurred, however, is less well documented. Such attempts are complicated because accidents are rare events and the data needed to assess effects are at best hard to come by.

Some research has been reported on the effects of accident experiences. Brown and Copeman (1975), for example, examined post-accident attitudes and beliefs and found that individuals who have been involved in several accidents tend to underestimate the consequences associated with accidents. Others have evaluated the effects of accidents on subsequent performance. One example here is that the use of near-miss information as feedback has been shown to reduce accidents in the forestry industry (Lagerlof, 1982). Another example is that feedback has been shown to improve driving performance (Lewin, 1982). These and a few other efforts notwithstanding, there is still very little research on the effects of people's accident experiences on subsequent accident events.

The research reported here explores patterns of accident involvement in a manufacturing complex for employees having at least one job-related injury during an 11-year period. The time between accidents for individuals was related to various circumstances associated with the events. In particular, intervals between consecutive accidents were considered in relation to factors specific to the individual and the environment as well as to factors having to do with the worker's activities and situation. The results of these analyses were then used to draw conclusions about the circumstances of an accident event that influence subsequent accidents.

4.2 METHOD

A database containing accident reports from a large petrochemical complex provided the context within which this work was carried out. The data were coded and computerized in a format that lends itself to the kinds of analyses of interest here. These analyses include defining a number of variables or parameters and examining relationships between them.

4.2.1 The database

The data consisted of information about accidents that occurred in the petrochemical complex over an 11-year period: 1 January 1981 through to 30 April 1992. Details of the database are presented elsewhere (Laughery *et al.*, 1983); a summary of its structure and content is included here.

An incident at this facility is considered an accident and gets into the database if an employee has to report to the first-aid station. The range of injuries may be very minor to very severe. Indeed, occasionally circumstances may occur in which someone reports to first aid and there is no injury involved. This situation may result from certain rules, but is very infrequent and not relevant to the analyses here.

When an accident occurs and an employee reports to the first-aid station, a report is filled out. This is a First Report of Injury or Illness, a fairly standard FRII. It is completed by at least two different people: a medical person, usually a nurse, and the employee's immediate supervisor. When an injury is more serious, as defined by certain criteria, an Occupational Safety and Health Administration (OSHA) report is also completed. From these documents, the FRII and the OSHA report (when relevant), information is extracted, coded and entered into the computerized database. The data consist of 20 coded variables about each accident. The variables can be characterized or divided into three categories: *demographic*, *injury* and *scenario*.

The demographic category consists of a number of facts that describe the circumstances in which the accident took place as well as some information about the injured employee. The employee data include identification of the person (badge number), gender and job classification. The location (place in the plant) and time (time of day, day of week, month and year) of the accident are also recorded. In short, this category represents the who, where and when types of information.

Injury data consist of three variables: type of injury, part of the body injured and severity of the injury. Injury type has 17 categories and includes such codes as contusion, laceration, sprain, burn, fracture, heat stress, etc. Body part injured has 16 categories. The injury severity code has six levels: no injury, minor injury, doctor case, restricted work, lost time and fatality. There were no fatalities at this facility during the 11-year period recorded. The definition of minor injury was any injury that did not require an OSHA report to be filed, that is it required first-aid treatment only.

The scenario category includes a set of six variables that represent the dynamics of the accident event itself, that is information about what happened. These codes break down the accident into a series of time-sequenced elements. If one thinks of an accident as happening over time, these codes partition the event into a series of subevents. Figure 4.1 shows the six scenario variables and one can think of the figure as a left-to-right time-line. Two of these variables, the prior activity and the accident event, are of particular interest here and will be described further. The others are discussed in detail by Laughery *et al.* (1983).

The scenario begins with the *prior activity*, which is the activity or task in which the worker was engaged just prior to the accident. Fourteen different activities make up the code for this variable. Table 4.1 shows the activities. The last item, 'not elsewhere classified', is a wastebasket code for activities that do not fit any of the other codes.

The *accident event* is the happening or circumstance that starts the accident sequence. In a sense, it is what happened at that point in time when events deviated from the normal or intended sequence and subsequently resulted in some sort of injury to an employee. Fourteen types of events constitute the code for this variable and are shown in Table 4.2.

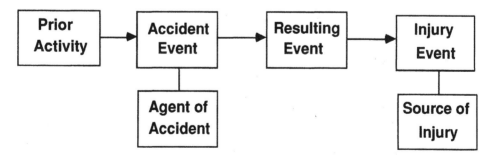

Figure 4.1 Accident scenario code.

Table 4.1   Prior activity code

Manually operating a valve
Taking a sample
In transit within work area
Changing elevations (stairs, ladders, etc.)
Assembling equipment
Disassembling equipment
Operating mobile equipment
Operating fixed equipment
Inspection or troubleshooting
Housekeeping
Handling materials
Responding to a hazardous situation
Away from work area
Not elsewhere classified

Table 4.2   Accident event code

Tool slipped on or missed the workpiece
Missed proper step
Lost balance
Slipped
Tripped
Overexertion
Non-containment of process materials
Equipment failed (other than primary containment)
Impact with object
Sudden movement of material
Excessive environmental stress (noise, heat, etc.)
Horseplay
Action by second person
Accident event unknown

The total number of accidents in the database was 11 322. Of these, 9425 were in the minor category and 1406 were major. Again, major refers to OSHA recordables and includes the doctor case, restricted work and lost time categories. The remaining 491 were accidents for which the data did not permit severity to be classified. The ratio of minor to major was 6.7 to 1.

Figure 4.2 shows the history of minor and major accidents at the facility over the 11-year period represented in the data. The number of minor injury accidents declined over this period. Some of this decline may have been due to a decline in the workforce, which reached its peak around 1981 or 1982 and dropped off significantly in the mid 1980s with the decline in the oil business at that time. During the 11-year period represented in the data, the number of employees at the facility varied between approximately 2500 and 3300. The number of major injury events stayed fairly constant over the 11 years, with the exception of 1986 and 1987, years that witnessed an increase. No clear explanation exists for the increase during this two-year period.

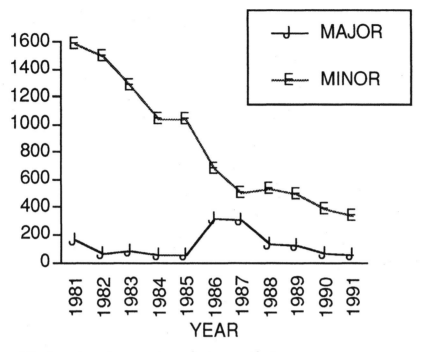

Figure 4.2 Annual number or major and minor accidents.

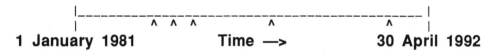

1 January 1981   **Time —>**   **30 April 1992**

Figure 4.3 Time-line of individual accident history.

4.2.2 The modified database

From the above database, a new or modified data set was created in a format that was appropriate for the kinds of analyses carried out in this study. This modified data set was organized by employee. For each employee a chronological listing of accidents was created with the time lapse between consecutive accidents recorded. For example, Figure 4.3 displays a time-line of the accident history for a worker who was involved in five accidents (denoted by the carets on the time-line). In this example, six episodes are formed. The first, from 1 January 1981 to the first caret, is left-censored and is not useful data because it is not known how much time had lapsed since the previous episode. The sixth episode, from the fifth caret to 30 April 1992, is right-censored and is not used in the various analyses reported here. The four between-caret episodes represent the kind of data on which our analyses are based. The data for each accident still include all of the demographic, scenario and injury codes.

A basic unit of analysis, or dependent measure, in this study was the time-lapse between successive accidents. A focus of our interests was the extent to which characteristics of successive accidents are the same or different and how these characteristics are related to the time-lapse. If circumstances characterizing the first accident, such as the task being performed, somehow enter into the learning experience of the employee, longer times between incidents involving the same task would be predicted.

4.2.3 Analytic procedures

A major goal in this work was to explore what factors influence how long an employee goes between accidents. We were especially interested in how characteristics of one accident influence how long it will be until the next one and what the characteristics of the next one are; specifically, are the characteristics similar? One of the analytical tools employed was regression analysis. More specifically, a Cox regression technique (Cox, 1972; Cox and Oakes, 1984) was used, which allows one to predict the probability that the next accident will occur in a given timeframe as a function of some predictor variable. This type of analysis essentially grew out of a class of statistical techniques that were used by industrial engineers and biomedical scientists to describe the useful lives of machines and the survival of patients. In the context of occupational health and safety, it can be used to explore questions such as: (1) how long do individuals remain accident-free following an injury they have had on the job and (2) does this time vary as a function of factors associated with the individual and/or the circumstances surrounding the previous accident?

4.3 RESULTS

Figure 4.4 shows the distribution of times between successive accidents collapsed across all employees and accident types. An exponential equation or curve fitted to this data is referred to as a survivor function.

Table 4.3 summarizes a few facts about the data. Over all employees there were 12 786 *episodes*, that is pairs of accidents. These episodes were generated by 2259 different people. The average number of episodes per employee was 5.7 over the 11-year period. The median time lapse between incidents was 10.5 months. The last three lines in the table provide a feel for the intervals that were bounded by two accidents. Fifty per cent of the intervals were longer (or shorter) than 18 months, while approximately 40% lasted more than 20 months. Slightly less than 18% were five years or longer.

Table 4.3 Summary statistics for episodes

2259 employees generated 12 786 episodes
Average number episodes per employee = 5.7 (SD = 4.6)
Median time lapse between incidents = 10.5 months
50% of the episodes lasted 18 months or longer
40% of the episodes extended more than 20 months
Less than 18% of the episodes lasted 5 years or longer

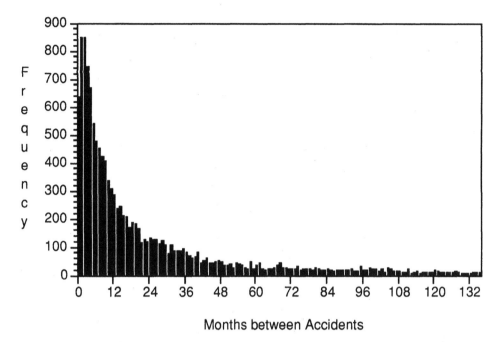

Figure 4.4 Distribution of times between consecutive accidents.

4.3.1 Factors influencing time intervals (episodes)

As noted, a major goal was to explore what factors influence how long an employee goes between accidents. The factors examined fall into several categories. The first can be labelled *historical*. Here the total number of accidents an employee had prior to the event beginning an episode and the duration of his or her previous episodes were examined. Not surprisingly, the more previous accidents and the shorter previous episodes, the shorter the time-lapse to the next one. This is not particularly interesting, however, because it simply says that people who have more accidents are more likely to have a shorter time-lapse for a particular episode. Similarly, the fact that *job category* as a factor showed longer episodes for incumbents of some jobs than for incumbents of others was not especially meaningful since the total number of accidents varies between jobs as a function of hazard level or risk (Laughery and Vaubel, 1993).

Temporal factors were also examined. Does the time of day, day of the week or month of the year of an incident influence how long it will be to the next one? Does the first one happening during overtime work matter? The answer to these questions is no.

Employee *gender* was difficult to analyze as a factor because of the limited number of accidents experienced by female employees (1268) as compared to male employees (10054). Generally, there was little difference in the duration of episodes that could not be accounted for by the fact that women had fewer accidents than men.

Three characteristics of the accident that begins an episode were explored as potentially influential on the duration of episodes as well as on the similarity of the

Table 4.4 Median number of months between accidents with the same or different prior activities (prior activity label refers to first accident)

Activity	Same	Different
Assembling/disassembling equipment	8.6 ($n = 1141$)	5.9 ($n = 877$)
Materials handling	16.0 ($n = 907$)	6.6 ($n = 958$)
In transit	28.9 ($n = 732$)	7.4 ($n = 649$)
Operating process equipment	22.7 ($n = 602$)	9.0 ($n = 504$)

two accident events that define an episode. These characteristics or factors were the prior activity and accident event variables in the scenario codes and the nature of the injury.

The first factor, prior activity, is the task or activity the employee was doing at the time of the two accidents. Is the time between incidents, the duration of the episode, longer or shorter when the tasks are the same, when they are different or does it matter? Four activities that constituted the prior activities most frequently involved in accidents were analyzed. These four are:

- *in transit* – walking, climbing stairs, ladders or scaffolding;
- *assembling/disassembling equipment* – taking apart or putting together equipment;
- *material handling* – handling or moving materials, housekeeping; and
- *operating process equipment* – manually turning valves, taking a sample.

Table 4.4 shows for each of the prior activities beginning an episode, the number of times the episode ended with the same or a different activity. These are the values of *n* in each of the cells. In terms of frequency, an accident involving a prior activity is more likely to be followed by an accident involving that same activity than one involving any of the other three activities. Indeed, except for materials handling, the 'same' activity frequencies were higher than all of the other 'different' activities combined. This outcome is not surprising since within given jobs a specific task would be likely to occur more frequently than others; thus, there would be more opportunity for the tasks to be the same.

However, when we look at the lapsed time between the two consecutive incidents, a different picture emerges. In all of the four prior activity circumstances, the times between accidents are longer when the two activities are the same. For three of the activities, the differences are substantial: materials handling, in transit and operating equipment.

A similar analysis was carried out for the accident event variable. The five most frequently occurring accident events were selected: slip/trip, tool manipulation, overexertion, person hit or contacted an object, and person hit by an object. The results of this analysis are shown in Table 4.5. Again, the number of same or different episodes are shown as the values of *n* for the various cells. With regard to episode duration, in all cases the times between incidents where the accident event codes are the same are significantly longer than the times for episodes where they are different.

The next analysis carried out on the episode data concerns back injuries. Since back injuries can be cumulative, one might expect that episodes between incidents of reported back injuries might be shorter than episodes where the first event

Table 4.5   Median number of months between accidents with the same or different accident events (accident event label refers to first accident)

Activity	Same	Different
Slip/trip	34.9 ($n = 329$)	7.9 ($n = 478$)
Tool manipulation	26.8 ($n = 335$)	7.2 ($n = 651$)
Overexertion	13.0 ($n = 508$)	7.5 ($n = 711$)
Person hit object	14.4 ($n = 1112$)	6.7 ($n = 1153$)
Object hit person	22.0 ($n = 1071$)	6.6 ($n = 1015$)

Table 4.6   Median number of months between same or different accidents based on back injuries

First accident	Second accident	Median
Back	Same	26.2 ($n = 276$)
Back	Different	8.2 ($n = 615$)

involved a back injury and the second event involved a different body part. On the other hand, if the employee learns from or alters his or her behaviour in an appropriate way on the basis of a back injury, then one might expect episodes of back-back injuries to be shorter than back-then-something-different injuries. Table 4.6 shows the frequencies of these two categories of events, as well as the time-lapse data. The back-back episodes have a time-lapse of more than three times that for the back-different episodes.

A final analysis concerned major and minor injuries. The question or issue of interest is whether the episode time following a major injury accident is longer than after a minor injury accident, regardless of whether the second event is major or minor. The number of episodes starting with major injury accidents was 1393 and the number starting with minor injuries was 9138. The median number of months worked following major injury accidents before having another accident was 13.6, while the duration following minor injury accidents was 9.3. Clearly, employees have longer accident-free times following major injury events.

4.4 DISCUSSION

Historical, job category, temporal and gender factors seem to be generally unrelated to time between accidents beyond their role in influencing the total number of accidents. What does seem to matter in how long an employee goes between two accidents are the characteristics of the first accident such as the activity or task the employee was performing, the event that precipitated the accident and the severity of the injury.

The outcomes of the prior activity and accident event analyses (Tables 4.4 and 4.5) indicate that the times between accidents where these circumstances are repeated are substantially longer than times between accidents where they are different. Certainly, these results are consistent with the notion that the employee's behaviour is modified by the prior accident experience. Also consistent with this notion is the outcome of the analysis showing longer episode durations following major injury accidents than following minor injury accidents.

There are, of course, many mechanisms or explanations that might be offered for the effects an accident experience has on an employee's future safety behaviour. Learning about the nature of hazards or outcomes may be one of the lessons of having an accident. Learning as a result of an accident about a chemical burn hazard associated with handling a particular chemical may lead to behaviour that lessens the probability of, and/or increases the time to, a similar event. In this same regard, the ever ubiquitous concept of attention may come into it in that the accident experience may influence what aspects of tasks or environments are allocated greater attention.

In addition to learning about hazards and learning how to respond to them, motivational factors may be another aspect of employee behaviour that is modified by accident experiences. If an employee allocates greater time and effort to addressing particular safety issues or hazards, longer times between accidents involving those hazards would be expected.

Supervisory control may also be relevant to explaining some of the results found here. Supervisor influences such as additional training (a learning effect) and rewards and punishments (a motivational effect) following an accident can affect the employee's behaviour.

Another type of accident outcome that could obviously affect subsequent accidents is job modifications or changes. In the analyses presented here an effort was made to control such factors by restricting the episodes examined to pairs of accidents where the employee was in the same job. Nevertheless, there may have been occasional circumstances where some modification took place after the first accident. For example, following a major injury accident or an accident involving a back injury, a period of time in which an employee was given light duty work assignments would have potentially lengthened the time to the next incident.

One of the interesting questions associated with the issues addressed in this work is how long the effects of having an accident last. Do the modifications in employee behaviour resulting from learning or motivation (or any other factor) dissipate and if so, how soon? Unfortunately, our database at this point does not permit us to explore this issue. Nevertheless, one of the lessons from this work is that there may be some significant payoffs in capitalizing on employees's accident experiences to modify their safety behaviour.

References

BROWN, I.D. and COPEMAN, A.K. (1975) Drivers' attitudes to the seriousness of road traffic offenses considered in relation to the design of sanctions. *Accident Analysis and Prevention*, **7**, 15–26.

COX, D.R. (1972) Regression models and life tables. *Journal of the Royal Statistical Society*, **34**, 187–200.

COX, D.R. and OAKES, D. (1984) *Analysis of survival data*, London: Chapman & Hall.

GREENWOOD, M. and WOODS, H.M. (1919) A report on the incidence of industrial accidents upon individuals with special reference to multiple accidents. In *Accident research*, eds W. Haddon, E.A. Suchman and D. Klien, New York: Harper & Row.

KOMAKI, J., HEINZMANN, A.T. and LAWSON, L. (1980) Effects of training and feedback: component analysis of a behavioral safety program. *Journal of Applied Psychology*, **65**, 261–270.

KOMAKI, J., COLLINS, R.L. and PENN, P. (1982) The role of performance antecedents and consequences in work motivation. *Journal of Applied Psychology*, **67**, 334–340.

LAGERLOF, E. (1982) Accident reduction in forestry through risk identification, risk consciousness and work organization change. *Proceedings of the 20th International Congress of Applied Psychology Edinburgh.*

LAUGHERY, K.R. and VAUBEL, K.P. (1993) Major and minor injuries at work: are the circumstances similar or different? *International Journal of Industrial Ergonomics*, **12**, 273–279.

LAUGHERY, K.R., PETREE, B.L., SCHMIDT, J.K., SCHWARTZ, D.R., WALSH, M.T. and IMIG, R.G. (1983) Scenario analyses of industrial accidents. *Proceedings of the System Safety Conference, Houston.*

LEWIN, I. (1982) Driver training: a perceptual motor skill approach. *Ergonomics*, **25**, 917–924.

MINTZ, A. (1954) Time intervals between accidents. *Journal of Applied Psychology*, **38**, 401–406.

(1956) A methodological note on time intervals between consecutive accidents. *Journal of Applied Psychology*, **40**, 189–191.

RASMUSSEN, J. (1985) Trends in human reliability analysis. *Ergonomics*, **28**, 1185–1195.

CHAPTER FIVE

Time-to-contact

PETER A. HANCOCK AND MICHAEL P. MANSER

5.1 INTRODUCTION

When one thinks of occupational injury, the mind's eye immediately conjures up a picture of accidents in the workplace. Typically, our vision encompasses a major event, usually located in an industry where the principal form of work involves considerable physical effort. A worker is lying stunned or unconscious on the ground having suffered severe trauma to a major body part. Emergency services are rendering aid and an investigation into the event is already in its beginning stages. A more contemporary vision of occupational injury might be set in the open-plan office. This time we see a worker not suffering from an acute injury but the victim of some form of repetitive strain trauma which makes continued computer-based data-entry work insupportable. Each of these visions is a valid view of the problems we try to address and solve. However, in this chapter, we want to put a third vision forward. This vision is framed in no single physical workplace, the worker is not amenable to even a general form of stereotyping and the work is itself highly diverse. The one constant across these situations is transportation. Since transportation workers occupy a mobile and frequently dangerous workplace, it is not difficult to envisage injuries as major concerns. Transportation injuries represent a significant and growing proportion of all occupational injuries. Furthermore, accidents that are confined to a specified workplace rarely affect non-workers or bystanders. In contrast, transportation-related accidents frequently affect individuals beyond the involved workers themselves. In consequence, transportation accidents can often have a much higher public profile. It is for these reasons that we want to consider the causes of transportation accidents and the technologies that are emerging which promise to alleviate their occurrence or at least mitigate their more harmful effects. To accomplish this, we are going to focus on one specific area of research with which we have direct familiarity, namely time-to-contact (Caird and Hancock, 1994; Manser and Hancock, 1996).

To illustrate time-to-contact, let us consider a specific example. Let us suppose that you are travelling on a twisting two-lane highway and have been unfortunate enough to be behind a slow, large truck for some extended period of time. On occasion, you have pulled out from behind the truck to ascertain whether a sufficiently straight portion of road is coming up so that you can pass. As you manoeuvre to view the road, you see a straight section ahead and in the distance there are two vehicles side by side. This could be two vehicles passing each other in opposite directions, in which case the vehicle in the left lane (the right lane in countries that drive on the left) is closing rapidly with you. It could be that the vehicle in the left lane is overtaking the other at a speed either faster or slower than your own. Consider this latter circumstance in comparison with the case of the oncoming vehicle. At a distance, both are changing their relative location and both are presenting an expanding visual outline at or very near the threshold of visual detection. It is on this minimal informational difference that you have to decide to pass or not to pass. The actual physical difference between these respective circumstances might represent a difference of some 200 feet per second in approach velocity. Some might argue that waiting a short time will disambiguate these or essentially resolve respective differences. However, in that short time the opportunity to pass might evaporate leaving a frustrated driver willing to tolerate a much more dangerous situation at the next possible 'opportunity'. This latter 'opportunity' might be the last one that driver ever takes. Under these circumstances visual information about time-to-contact can be the basis, literally, for a life or death decision.

In what follows, we look at the philosophical and historical antecedents of such time-to-contact issues and then provide specific examples of contemporary time-to-contact research findings. We then examine how the results of these investigations can be used to inform designers of collision-avoidance systems as envisaged in recent Intelligent Transportation Systems (ITS) technical innovations (Hancock *et al.*, 1993). Finally, we illustrate how time-to-contact can be used beyond the transportation realm as a general construct to address injury prevention in a wide variety of occupational pursuits.

5.2 HISTORICAL AND PHILOSOPHICAL BASIS

For those whose concern is solely for the practice of occupational injury prevention, we beg indulgence. For it is in this section where we seek to provide at least a brief overview of the philosophical and historical basis of time-to-contact concerns and the reasons why such a construct is of importance. However, we are not such committed academics that we do not recognize the immediacy of the demands of the real world. Therefore, the practically oriented reader might wish to scan briefly over the present section to reach the subsequent ones, which deal with applied issues more directly. However, for the more general reader, the following should provide at least a sufficient outline to trace the philosophical origins upon which we base our work.

One of the great philosophical contentions of human knowledge concerns the argument as to what degree any individual can 'directly' experience the world around them. For many centuries, the accepted position was that the world was experienced very indirectly reflected as 'shadows on a cave wall' (Plato, see Hare, 1982). The rationalists and empiricists also accepted that what we experience are 'ideas' of the world, filtered only dimly and indirectly through the inefficient and selective senses. One philosopher, applying Occam's razor to the question of perception, dismissed physical matter altogether as a causal agent for sensation, relying on the omnipotence of God to supply such ideas 'directly' to the mind of each individual (Berkely, see Luce and Jessop, 1948). What nullified this latter proposition was not the supremacy of matter but a growing scientific doubt in the existence of such a Deity. Despite such steps, the history of psychology as a formal discipline reflects the continuing belief that a poor and impoverished 'image' of the world is provided by the senses (Todd, 1981); an image which is augmented, completed and subsequently interpreted by the cognitive abilities of the individual. This position has never lacked its critics. One fundamental flaw of this traditional position, which can be labelled the 'constructionist' view (since the perceived world has to be constructed from component sensory data), is the role of evolution. Today, we recognize that the very sensory systems that are the subject of dispute grew and evolved in a symbiotic relationship with the 'world' that was to be perceived. Hence, the relationship between the specific sensory system and what is to be sensed is a much more intimate and therefore direct one than early philosophers conceived.

Recent decades have seen the growth of 'ecological' psychology, which has emphasized this intimate relationship (Gibson, 1979). Ecological psychology also raises awkward points for the traditional contructionist view, particularly the question as to how the finite capabilities of the central nervous system can cope with the potentially infinite possibilities of an unknown environment. The fundamental contention of the ecological approach is that, to a degree, no manipulational, translational or 'constructional' mechanisms are required of the central nervous system since objects that occupy a world are perceived 'directly'. Arguing from a position of parsimony, such a view emphasizes the elegance of this form of perception. Elsewhere the argument that the fundamental nature of the question of 'directness' is itself flawed is presented (Hancock, 1993). For the present purposes, however, it is important to examine the evidence offered in support of the ecological position. One central tenet of the ecological position concerns the perception of approaching objects. This general area is known as 'time-to-contact'. The crucial argument in this area concerns a phenomenon known as *tau*, which as a rate of optic expansion, a concept we discuss at length below, purports to specify time-to-contact for an individual.

The individual who dismissed the need for matter was the same one who inquired how it was that we are able to perceive depth since such a line was projected end-on into the 'fund' of the eye (Berkley, see Luce and Jessop, 1948). One form of answer to this question is that motion can 'disambiguate' the situation either by lateral movement of the head and eyes, revealing further information about other 'faces' of an object of concern, or by approaching/retreating from an object or having that object approach or retreat from us. Should an object approach with symmetrical expansion about any axis, it provides a vital survival signal since it specifies that this object is on a potential collision trajectory. Thus, this 'time-to-contact' recognition is vital for any organism hoping to survive the vagaries of an uncertain and predator-laiden environment. If any capability should be subject to 'direct' perception then, surely the one specifying time-to-contact is the pre-eminent candidate.

5.3 THE OPTIC ARRAY

To explore issues related to time-to-contact it is necessary to first define the difference between stimulation and stimulus information (Gibson, 1979). Gibson advanced two seemingly paradoxical statements to aid our understanding of visual perception. According to Gibson nothing can be seen but light, but paradoxically light can never be seen. First, the only thing that enters the eye is light in the form of wave particles or rays. These light waves/rays are projected from a luminous object or reflected from object surfaces in the environment. These light rays travel to the eye, enter and are projected to the back of the eye. It is on the retina that light waves/rays stimulate photoreceptors so that observers perceive their environment. Second, the contention that humans cannot see light is also true when we consider the fact that light waves or rays do not have structure, matter or weight and as a consequence cannot be seen. Seeing is a process of photoreceptor stimulation. When humans see a laser beam what is actually being perceived is the reflection of light energy off molecules residing in the air. When humans see an approaching vehicle we do not actually see the object, but instead what is perceived is stimulation of photoreceptors on the retina caused by the reflection of patterns of specific light energy off the surface of the approaching vehicle.

It is these patterns of light energy that are projected to the retina that are termed the 'optic array' by ecological psychologists examining issues in visual perception (Gibson, 1979). The optic array is not static but rather is a dynamic field and represents a directly perceived optical flow field. *Global optical flow* field patterns are one type of flow field. Global optical flow field patterns occur when the entire optical array of light is moving on the retina. For example, when a person is moving through an environment, global optical flow field patterns are generated on the back of the retina and are continually flowing off the edges of the retina in all directions. As a person moves through an environment, images appear directly in front of them in their foveal vision while images gradually are removed from sight around the periphery of the visual field. Global optical flow field patterns can also be produced on the retina while a person is moving backwards through an environment. In this case images are flowing across the edges of the retina towards the retinal centre.

A second type of optical flow field pattern has been labelled *local optical flow.* Local optical flow field patterns are characterized by discrete light patterns that change shape, position or size on the retina. Local optical flow field patterns are experienced when a person is stationary and is watching an object approaching their position. Under these conditions only a portion of the entire optic array is moving. These different categories of flow field motion can be further subdivided as either *lamellar* or *radial*. If a person is stationary and looking forward and an object travels from left to right in front of them, this is a *lamellar optical flow* field. *Radial optical flow* occurs if the object is approaching on a head-on collision course. In this situation the image is stationary on the retina expanding in all directions. Global optical flow field patterns and local optical flow field patterns might appear to be dichotomous, however, they are actually ends of a continuum. For example, often in driving the optical array contains localized optical flow fields embedded within a global optical flow field, such as when watching a car approaching in the opposite lane while driving forward yourself. Also, in driving around curves the optical flow field of the far point of the road is in lamellar flow, but the optical flow field immediately in front of the driver is radial in nature. An important question is how this compound optic flow relates to time-to-contact. However, before attempting to answer this question we need to clarify each of the various terms that have been used in relation to time-to-contact in general.

5.4 A TIME-TO-CONTACT TAXONOMY

One of the sources of confusion concerning time-to-contact is the different terms that have been used in relation to the general phenomena. Apparently there have been no explicit attempts to clarify this confusion and to provide a formal taxonomy for the different labels that have been attributed to this phenomenon. Each of the following terms have been employed: *arrival time* (Schiff and Oldak, 1990; DeLucia, 1991; Caird and Hancock, 1994), *time-to-arrival* (Schiff and Oldak, 1990), *time-to-coincidence* (Groeger and Brown, 1988; Groeger and Cavallo, 1991; Groeger *et al.*, 1991), *time-to-collision* (Purdy, 1958; Schiff, 1965; McLeod and Ross, 1983; Brown and McFaddon, 1986; Cavallo *et al.*, 1986; Cavallo and Laurent, 1988; Tenkink and Van der Horst, 1990; Groeger and Cavallo, 1991), *time-to-contact* (Lee, 1976; Tresilian, 1991), *time-to-go* (Carel, 1961) and *time-to-passage* (Kaiser and Mowafy, 1993). At first it might appear that these terms have been used generally to represent the ability to estimate when a moving object will reach a second object or observer in space. However, this has not always been the case. Here, we provide a definitive taxonomy that describes what is meant by each of these terms in detail. First, we show a diagrammatic representation of the conditions that compose this taxonomy

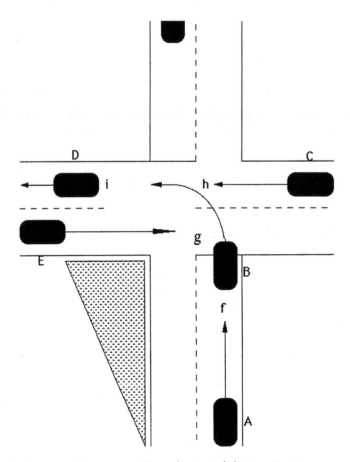

Figure 5.1   A diagrammatic representation of a typical driving situation.

(Figure 5.1) and finally we provide a real-world example where each of these facets of the general time-to-contact realm exert their specific influence. Figure 5.1 depicts a traffic intersection. Although the vehicles in this depiction are being 'driven' on the right-hand side of the road it is obvious that comparable situations occur when the vehicles are driven on the left-hand side of the road.

(i) *Time-to-contact:* The term time-to-contact indicates that a stationary observer (in vehicle B) views a vehicle (E) approaching them on a collision course. The traditional research approach to such conditions depicts a vehicle approaching the observer and, while en route to that observer, the approaching vehicle disappears from the scene. The observer has to indicate, typically via a button press, when the approaching vehicle would have reached their position had it not disappeared. In this situation a radially expanding optical flow field pattern is generated on the back of the retina.

(ii) *Time-to-passage:* Similar to time-to-contact, time-to-passage describes a stationary observer (vehicle B) viewing an approaching vehicle (C) and at some point during the vehicle's approach the vehicle disappears from the driving environment. However, unlike time-to-contact, the approaching vehicle is not on a direct collision course with the observer. Rather if the approaching vehicle had not disappeared from the scene it would have passed just in front of the observer. Again, the observer's task is to respond when they felt the approaching vehicle would have passed in front of their position. In this situation the type of optical flow field is local and the pattern is primarily radial in nature when the vehicle is at a distance. However, when the vehicle is close to the observer the flow field pattern becomes lamellar in nature.

(iii) *Time-to-go:* Time-to-go is similar to time-to-contact and time-to-passage in that a stationary observer (vehicle B) is viewing an approaching vehicle (E). In time-to-go the approaching vehicle can be on a collision course or a by-pass course for the observer and the approaching vehicle can disappear en route to the observer or can travel the entire distance to the observer. The observer is required to respond when they feel they should move, or more accurately accelerate, in order to avoid collision with the approaching vehicle. In Figure 5.1, this would mean that vehicle B would have to drive across the intersection and make a left-hand turn before a collision with vehicle E occurred. The optical flow field patterns in this situation are identical to those in time-to-passage.

(iv) *Time-to-arrival (arrival time):* We view the terms time-to-arrival and arrival time as synonymous. In time-to-arrival a moving observer (vehicle A) is approaching a stationary target (vehicle B) or point in space (point f). At some point during the observer's approach to the stationary target or position in space the scene becomes blank or the observer's vision is obstructed. It is the observer's task to estimate when they would have reached the predetermined target or position in space had the scene not gone blank or had their vision not been obstructed. In time-to-arrival a global optical flow field is generated on the retina.

(v) *Time-to-collision:* Time-to-collision involves having a moving observer estimate when they will collide with another moving object. In Figure 5.1 the moving observer (in vehicle B) is required to determine when they will collide with vehicle C at location h. In this scenario, the optical field pattern for the observer consists of localized flow embedded within a global optical flow field pattern.

(vi) *Time-to-coincidence:* Time-to-coincidence refers to a collision between a moving object (vehicle E) and a stationary object (vehicle B) which another individual (in vehicle A) observes from a distance. In this configuration one or both moving objects disappear before coinciding with the other object. The participant's task is to indicate when the moment of coincidence between the two objects would have occurred. In this scenario there are two localized optical flow field patterns generated on the retina.

(vii) *Unperceived time-to-contact:* There is a final form of contact perception, which is the structural and functional failure to perceive the approaching object. Obviously, under any of the conditions we have indicated, it is possible for an individual to have objects within their visual field and yet, for a number of functional reasons connected to neuropsychological and neurophysiological processes, fail to register and respond to an object's presence. There are extensive questions associated with these 'higher' level functions that we have not addressed here. However, there is an even more simple failure which we cannot pass by without comment: an individual may well be struck by an object that they never perceived. In visual terms this might mean an object approaching from the rear (with no rear mirror) or approaching in the blind spot. We should, however, note that auditory time-to-contact might prove of use in alerting an individual in this situation. We have labelled this *unperceived time-to-contact* and used the global lamellar case as one exemplar. In many occupational accidents caused by collision, not having seen the object, or vehicle, causes a considerable percentage of events.

At this juncture, it is our purpose to relate what the observer perceives (that is the flow field characteristics) with the environmental configurations used in the various research situations presented above. As can be seen in Figure 5.2, we divide flow field characteristics into global and local, with radial and lamellar components under each.

We appreciate that the real world often presents complex combinations of these characteristics, however, one initial taxonomic differentiation is based on these four divisions. As can be seen from Figure 5.2, we have located each environmental configuration within its predominate flow field. There are some which do not fit precisely into this categorization. However, the taxonomy identifies one form of flow field not investigated under this regimen, that is a global, lamellar condition. We can imagine a number of situations in which this occurs, e.g. looking out of a moving train window or glancing over to another competitor at the end of a sprint race. Without a specific environmental reference, the closest comparable condition is time-to-passage.

In the present taxonomy, we have tried to include all basic conditions identified by contemporary researchers. In evaluating time-to-contact type scenarios, it is clear that initial experimental effects have focused on a restricted number of relatively simple situations and drawn on dichotomies of whether the observer or the environment represents the major source of movement. However, it is also clear that in the real world, when observers either move or are stationary, objects in the flow field also approach and recede in complex patterns. Therefore, our present approach represents a working framework rather than an exhaustive description.

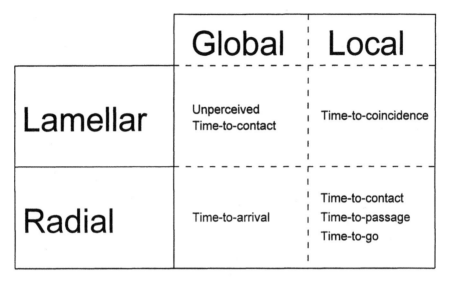

Figure 5.2   Taxonomy of optical flow field characteristics.

5.5 TIME-TO-CONTACT: CONTEMPORARY RESEARCH

As indicated in the previous section, to estimate time-to-contact researchers have used situations in which an object, typically a vehicle, is approaching a participant on a collision course and at some point en route the approaching object vanishes from the scene. The participant's task is to press a button when they felt the approaching object would have reached their position. This experimental technique has been labelled the 'removal paradigm' (Manser and Hancock, 1996). Results of experiments using the removal paradigm have indicated that there are several external and internal factors which influence the ability to estimate time-to-contact accurately.

One of the exogenous factors influencing estimates of time-to-contact is the velocity of the approaching vehicle. Specifically, when a vehicle is approaching a participant at higher velocities estimates of time-to-contact are more accurate than when the vehicle approached at lower velocities (McLeod and Ross, 1983; Schiff *et al.*, 1992). A second external factor influencing estimates of time-to-contact is the period of time a participant is allowed to view an approaching vehicle. Results indicate that when observers were allowed to view the approaching vehicle for longer periods of time, estimates of time-to-contact became more accurate (McLeod and Ross, 1983; Schiff and Oldak, 1990; Caird and Hancock, 1994). A third external factor related closely to viewing time is viewing distance. Research has indicated that participants' estimates of time-to-contact were more accurate when participants were allowed to see the vehicle for greater approach distances (Tresilian, 1991). This effect remained even when the viewing time was held constant and total viewing distance was manipulated.

In addition to the external factors influencing estimates of time-to-contact there are several internal factors. One of these internal variables is the sex of the observer (McLeod and Ross, 1983; Schiff *et al.*, 1992; Caird and Hancock, 1994). It has been found that males were more accurate than females in estimating time-to-contact as actual time-to-contact increased. In addition, it has been found that males' estimates of time-to-contact were significantly less variable than females' estimates of time-to-contact (McLeod and Ross, 1983; Schiff and Oldak, 1990; Caird and Hancock, 1994). Only one study has reported no difference for the sex of the observer (Schiff et al., 1992). Our recent work has indicated that the presence of a sex effect is contingent on the precise nature of the kinematic conditions under consideration (Manser and Hancock, 1996). Another internal influence affecting estimates of time-to-contact is the inherent limitations and capabilities of the human visual system. Manser and Hancock (1996) have shown that participants are more accurate when the vehicle approaches from a head-on collision course (an approach directed towards the front of the participant) as opposed to alternative angles of incidence (e.g. an approach directed towards the side of the participant).

One of the persistent characteristics in time-to-contact studies is the tendency to underestimate, progressively, time-to-contact as actual time-to-contact increases. This phenomenon permeates the research database as far back as 1958 when Knowles and Carel examined whether participants could determine the amount of time before a head-on collision would occur in the absence of familiar environmental cues such as size, distance and speed. One of their findings was that participants could determine time-to-contact fairly accurately up to about 4 s, beyond that estimates of time-to-contact were underestimated progressively. Later, Carel (1961) reported that estimates of time-to-contact were underestimated progressively as actual time-to-contact increased. The data from Carel's study were fitted with a straight line, which resulted in a slope of 0.74. More recently, Schiff and Detwiler (1979) used film footage to display a vehicle approaching on a head-on collision course in an effort to examine the effects of vehicle approach velocity and vehicle viewing distance. The results of their studies indicated that estimates of time-to-contact were underestimated at roughly 60% of actual time-to-contact. Similar underestimations were found by Schiff and Oldak (1990), Schiff et al. (1992) and Caird and Hancock (1994). In particular, in Caird and Hancock's experiment one of four vehicles approached a stationary participant on a collision course and was removed from the driving scene at one of two distances. The results of their experiment indicated that participants' estimates of time-to-contact were similar to other studies, with a slope of 0.56. Recently, Manser and Hancock (1996) investigated the effects of vehicle approach trajectory and vehicle approach velocity on estimates of time-to-collision. Our results also indicated that participants underestimated time-to-collision progressively as actual time-to-collision increased.

McLeod and Ross (1983) examined the effects of viewing time on estimations of time-to-collision using a slightly different research technique: their experimental film segments depicted participants travelling towards a stationary vehicle. While travelling towards the stationary vehicle the film segment went blank and the participant asked to respond when they felt they would have collided with the stationary vehicle. McLeod and Ross found that participants underestimated time-to-collision at approximately 60% of actual time-to-collision. When these data are fitted to a straight line the slope is 0.58. Using a similar experimental technique, Cavallo and Laurent (1988) had participants travel as passengers in a vehicle which was approaching a stationary object. Cavallo and Laurent examined the effect of driver experience levels, distance evaluations and vehicle approach speeds on estimates of time-to-collision. The results of their study indicated that estimates of time-to-collision were systematically underestimated. Similar to previously presented studies, when the data were fitted to a straight line the slopes were 0.73 and 0.57 for experienced and beginner drivers, respectively. Other experiments have revealed the tendency to underestimate time-to-arrival (Kaiser and Mowafy, 1993). Figure 5.3 shows a depiction of the slopes indicating this propensity to underestimate time-to-contact. It should be noted that recently Cavallo (personal communication) has divided results from previous research into two categories based on whether the participant is stationary, estimating when an object will reach their position, or whether the participant is moving through the environment estimating when they will reach a particular point in space. Cavallo found that the regression lines for participants who were stationary and estimating when a dynamic object would reach their position overestimated time-to-contact up to about 1.5 s. After 1.5 s participants began to underestimate time-to-contact progressively. Interestingly, the regression line for participants who were moving through an environment showed that they never overestimated time-to-arrival, but consistently underestimated it. There are several potential reasons for the differences in slope values between the two conditions and these are reflected in part in the taxonomic structure we have proposed.

Several authors have suggested underlying reasons for these persistent underestimations. Schiff and Oldak (1990) suggest that participant underestimations are due to a biological tendency to err on the side of safety to protect oneself from collision or contact in a potentially dangerous situation. From an ecological

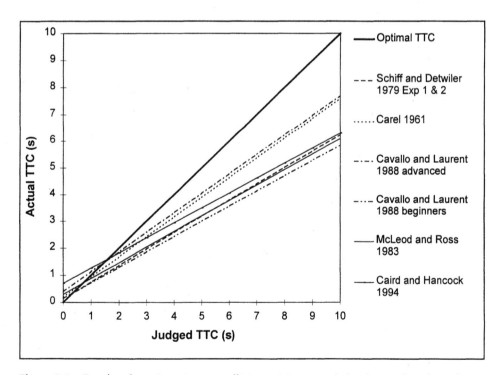

Figure 5.3 Results of previous time-to-collision (TTC) research for the cited studies. The collective findings show that participants underestimate time-to-collision and that such underestimation grows with the absolute duration of actual time-to-collision.

approach in visual perception it would seem reasonable that humans would have developed, over millennia, the tendency to underestimate time-to-contact in order to enhance the chance of avoiding a dangerous collision. Kaiser and Mowafy (1993) suggest underestimations may be due to a distortion of the visual/temporal space or that they could be a by-product of the cognitive extrapolations required of participants.

Although the explanations presented above have merit, a simpler alternative explanation is possible. This alternative explanation relates to the manner in which visually specified information has been used by humans throughout their evolution. In naturally occurring situations, when an object is approaching an observer, the approaching object travels the entire distance to the person or becomes occluded by another object while en route to the person. These naturally occurring situations are quite different from the traditional removal paradigm which has been used to examine various time-to-contact issues. Consequently, the degree to which people underestimate time-to-contact may well be an undesirable by-product of a fallacious research approach. Specifically, sudden vehicle disappearance in the removal research paradigm is a visual anomaly that humans have not been exposed to and have not learned to adapt to throughout their evolution. Our most recent investigation (Hancock and Manser, 1996) studied vehicle approach velocity, participant gender and participant age under an 'occlusion' condition as compared with the 'removal' condition. The results indicated that participants' estimates of time-to-contact were significantly more accurate when they viewed the former, more ecologically valid, research paradigm. These data confirm that participants are more accurate at estimating time-to-contact than the previous consensus indicates. Figure 5.4 depicts an illustration of the research conditions used by Hancock and Manser (1996).

Having given a brief overview of some of the theoretical issues associated with time-to-contact, let us now turn our attention to some applications. First and most prominently, the results of the study by Hancock and Manser (1996) confirm that the way a research question is posed has a significant impact on the results obtained in that research. In this case, the dependent measure in each research paradigm was identical (estimates of time-to-contact), however, by making a small change in the way the approaching vehicle was removed from the driving environment large and systematic differences occurred in the dependent response.

Second, it is necessary to pose research questions in the most ecologically valid manner possible so that the results may be maximally generalizable. One method for increasing the generalizability of research results is to display the simulated world in the most realistic manner possible. However, Kantowitz (1992) warns that an increase in physical fidelity between the real world and the simulated world does not necessarily enhance generalizability. What does enhance generalizability is that the psychological processes engendered by the real world and the simulated world are comparable. We agree with this contention, but point out that very little evidence has been produced which indicates what exactly are the essential psychological processes in the real world, and in particular the realm of driving, that are critical for replication or simulation. Hancock and Manser (1996) have provided evidence that one of the key psychological processes in the real world occurs when a vehicle becomes occluded by some object in the driving environment. This leads to questions concerning research approaches. First, are there other psychological processes

Figure 5.4   Depiction of the driving scenario used in Manser and Hancock (1996). Note that on the road approaching from the left there is a bush that serves to occlude an approaching vehicle. Note also that this is only an approximation of the actual scene. Limitations are inherent in viewing three-dimensional surfaces in a two-dimensional representation. Photograph courtesy of Neil Kveberg.

occurring in the real world which experimenters are ignoring in experimental research? The answer appears to be in the affirmative. Second, by ignoring a spectrum of potential influences, are previous data problematically confounded?

Although the results of previous studies are not totally generalizable to the real world due to the lack of a fully ecologically valid approach, there remain some important implications for real-world applications. To review briefly, studies using the traditional removal time-to-contact research paradigm and the more ecologically valid occlusion time-to-contact research paradigm have indicated that participants are more accurate in estimating time-to-contact when the approaching vehicle is viewed for greater distances, more accurate when the approaching vehicle is viewed for a greater period of time and more accurate when the approaching vehicle is approaching at higher velocities. Clearly, these results should be considered by traffic engineers who design roadways and intersections. Specifically, roadway designers should attempt to maximize viewing time and viewing distance at intersections, particularly unregulated intersections. The issue of increasing vehicle approach speed, although helpful in the laboratory for increasing the accuracy of estimates of time-to-contact, may not be a feasible alternative for real-world accident reduction tactics because of the possibility for greater damage and life lost at higher approach speeds. A second possible application of the results of time-to-contact studies is in intelligent transportation systems (ITS), formally known as intelligent vehicle highway systems (IVHS). New ITS applications include collision avoidance systems whose purpose is to prevent people from getting into vehicular accidents by taking charge of the vehicle in potentially dangerous situations. However, the variables used in the system algorithm to determine the potentially dangerous situations must come from somewhere and could be research results from time-to-contact studies. For example, collision avoidance systems would need the capability of calculating how well drivers could estimate time-to-contact under a variety of conditions to be able to determine at a particular moment if the drivers are capable of avoiding an imminent collision.

5.6 TIME-TO-CONTACT AND OCCUPATIONAL INJURY

We have framed all of our discussion about time-to-contact within a transportation-related realm. However, this is a very restricted view. As indicated in the opening of this chapter, time-to-contact is an absolutely vital capability for any organism hoping to survive in any environment. Therefore, the capability to perceive time-to-contact is a general one, as is its realm of application. We can well imagine any number of situations in which the ability to distinguish time-to-contact is critical in injury avoidance. Consider, for example, the construction worker: falling objects are of critical concern and a major form of injury causation. The ability to distinguish symmetrically expanding objects, that is those which will collide with the individual, from those with asymmetrical expansion, which will not, is a critical characteristic in avoidance strategy. Similarly, in semi-automated and automated manufacturing facilities the ability to distinguish potential collision courses by automated vehicles is vital in avoiding robotics-related accidents (Hamilton and Hancock, 1986). In fact, in all cases where objects collide with workers, time-to-contact specifies critical avoidance information. As we have illustrated for transport systems, such information can be augmented by the use of technical support systems. Consequently, we submit that such support systems can be of assistance in multiple realms of occupational injury prevention. It is not only single-collision events that can benefit from time-to-connect knowledge. Repetitive strain trauma, from keyboard use for example, is frequently attributed to posture while typing. However, this is to neglect the design of the keyboard and its characteristics in terms of pressure required to depress and operate keys. This, after all, is the primary task of keyboarding and this process is simply an extension of time-to-connect into a region that we call 'soft-collision.' In consequence, we believe that time-to-contact is a critical construct in battling the adverse effects of occupational injury and can add both a practical and conceptual tool to the professional's armoury in the never-ending fight to combat accidents and damage to individuals in the workplace.

References

BROWN, I.D. and McFADDON, S.M. (1986) Display parameters for driver control of vehicles using indirect viewing. In *Vision in vehicles*, eds A.G. Gale, M.H. Freeman, C.M. Haslegrave, P. Smith and S.P. Taylor, pp. 265–274, Amsterdam: North Holland.

CAIRD, J.K. and HANCOCK, P.A. (1994) The perception of arrival time for different oncoming vehicles at an intersection. *Ecological Psychology*, **6**, 83–109.

CAREL, W.L. (1961) *Visual factors in the contact analog*, Publication No. R61ELC60.34, Ithaca, NY: General Electric Company Advanced Electronics Center.

CAVALLO, V. and LAURENT, M. (1988) Visual information and skill level in time-to-collision estimation. *Perception*, **17**, 623–632.

CAVALLO, V., LAYA, O. and LAURENT, M. (1986) The estimation of time-to-collision as a function of visual stimulation. In *Vision in vehicles*, eds A.G. Gale, M.H. Freeman, C.M. Haslegrave, P. Smith and S.P. Taylor, pp. 179–183, Amsterdam: North Holland.

DeLUCIA, P.R. (1991) Pictorial and motion-based information for depth perception. *Journal of Experimental Psychology: Human Perception and Performance*, **17**(3), 738–748.

GIBSON, J.J. (1979) *The ecological approach to visual perception*, Boston: Houghton-Mifflin.

GROEGER, J.A. and BROWN, I.D. (1988) Motion perception is not direct with indirect viewing systems. In *Vision in vehicles II*, eds A.G. Gale, M. Freeman, C. Haselgrave, P. Smith and S. Taylor, pp. 27–34, Amsterdam: North-Holland.

GROEGER, J.A. and CAVALLO, V. (1991) Judgments of time to collision and time to coincidence. In *Vision in vehicles III*, eds A.G. Gale, I.D. Brown, C.M. Haslegrave, I. Moorehead and S. Taylor, pp. 27–34, Amsterdam: North-Holland.

GROEGER, J.A., GRANDE, G. and BROWN, I.D. (1991) Accuracy and safety: Effects of different training procedures on a time-to-coincidence task. In *Vision in vehicles III*, eds A.G. Gale, I.D. Brown, C.M. Haslegrave, I. Moorehead and S. Taylor, pp. 35–43, Amsterdam: North-Holland.

HAMILTON, J.E. and HANCOCK, P.A. (1986) Robotics safety: Exclusion guarding for industrial operations. *Journal of Occupational Accidents*, **8**, 69–78.

HANCOCK, P.A. (1993) The future of hybrid human–machine systems. In *Verification and validation of complex systems: Human factors issues*, eds J.A. Wise, V.D. Hopkin and P. Stager, Berlin: Springer-Verlag.

HANCOCK, P.A., DEWING, W.L. and PARASURAMAN, R. (1993) A driver-centered system architecture for intelligent-vehicle highway systems. *Ergnomics in Design*, **2**, 12–15, 35–39.

HARE, R.M. (1982) *Plato*, Oxford: Oxford University Press.

KAISER, M.K. and MOWAFY, L. (1993) Optical specification of time to passage: Observers' sensitivity to global tau. *Journal of Experimental Psychology: Human Perception and Performance*, **19**, 1028–1040.

KANTOWITZ, B.H. (1992) Selecting measures for human factors research. *Human Factors*, **34**, 387–398.

KNOWLES, W.B. and CAREL, W.L. (1958) Estimating time-to-collision. *American Psychologist*, **13**, 405–406.

LEE, D.N. (1976) A theory of visual control of braking based on information about time-to-collision. *Perception*, **5**, 437–459.

LUCE, A.A. and JESSOP, T.F. (1948) *The works of George Berkely: Bishop of Cloyne*, Nelson and Sons: Edinburgh.

MANSER, M.P. and HANCOCK, P.A. (1996) The influence of approach angle on estimates of time-to-collision. *Ecological Psychology*, **8**, 71–99.

McLEOD, R.W. and ROSS, H.E. (1983) Optic flow and cognitive factors in time-to-collision estimates. *Perception*, **12**, 417–423.

PURDY, W. (1958) The hypothesis of psychophysical correspondence in space perception. Unpublished doctoral dissertation, Cornell University, Ithaca, New York.

SCHIFF, W. (1965) Perception of impending collision: a study of visually directed avoidance behavior. *Psychological Monographs: General and Applied*, **79** (whole no. 604).

SCHIFF, W. and DETWILER, M. (1979) Information used in judging impending collisions. *Perception*, **8**, 647–658.

SCHIFF, W. and OLDAK, R. (1990) Accuracy of judging time-to-arrival: Effects of modality, trajectory and sex. *Journal of Experimental Psychology: Human Perception and Performance*, **16**, 303–316.

SCHIFF, W., OLDAK, R. and SHAH, V. (1992) Aging persons' estimates of vehicular motion. *Psychology and Aging*, **7**, 518–525.

TENKINK, E. and VAN DER HORST, R. (1990) Car driver behavior at flashing light railroad grade crossings. *Accident Analysis and Prevention*, **22**, 229–239.

TODD, J.T. (1981) Visual information about moving objects. *Journal of Experimental Psychology: Human Perception and Performance*, **7**, 795–810.

TRESILIAN, J.R. (1991) Empirical and theoretical issues in the perception of time-to-contact. *Journal of Experimental Psychology: Human Perception and Performance*, **17**, 865–876.

CHAPTER SIX

The use of human error data as indicators of changes in work performance

CLARE M. POLLOCK

6.1 INTRODUCTION

The development of safe working environments relies on the accurate identification of those factors that can contribute towards an increased risk of accidents and injuries at work. Similarly, it requires an understanding of the effect that changing the work environment can have on safety. These two goals have traditionally been achieved by using accidents or injuries at work as the indicator of unsafe performance. An environment is considered to be unsafe if it can be found to have caused, or contributed to, the incidence of accidents and injuries; a change to a work environment can be said to be safe if it results in no increase in accidents or injuries. This approach has intuitive appeal in that the ultimate goal of any assessment of occupational safety must be to reduce accidents and injuries. However, accidents and injuries may be insufficient indicators of the safety of a work environment or change to that environment due to the subtle or complex effects some work changes can have on safety. (Note, in this paper the terms 'accidents' and 'injuries' are used synonymously for convenience. It is recognized that not all accidents result in injuries and not all injuries are the result of specific accidents.)

When a workplace is altered, for example due to the introduction of a new system of work or equipment, the resulting effect could be to create both good and bad influences on performance, which will consequently tend to change the safety of the workplace. Some aspects of a change may improve performance but may also cause stressors that could decrease performance. Under such circumstances, the overall injury or accident rates may not change but the *types* of accidents or injuries that result could be affected, with consequences for necessary remediation actions.

An unchanged accident rate may not imply unchanged performance. For example, the general move to automate manual work has reduced much of the physical stress of manual handling and would be expected to have reduced the incidence of manual-handling injuries. However, the replacement of the manual tasks with tasks involving substantial computer and keyboard use might result in a higher incidence of work-related neck and upper limb disorders. In addition, the loss of the skill and practice required to perform manual tasks when they do occur may result in injuries from performance of the unfamiliar task. Overall, there may be no change in injury rate but the change in pattern of injury will highlight the need to address newly developing problems.

6.2 TWELVE-HOUR SHIFTWORK

The example that is used in detail in this chapter relates to the introduction of twelve-hour shifts in the workplace. When shifts of twelve hours' duration are introduced into a workplace in place of eight-hour shifts the (average) hours worked each week usually remains the same but the number of shifts is reduced. The twelve-hour shift system requires workers to work for longer when they are at work but provides more full days off. The shift pattern is frequently referred to as the compressed work week; a term that is used in this chapter. It is important to clarify that the compressed work week only includes those systems of working that set twelve hours (or ten hours in some organizations) as the standard shift length and have longer periods free from work to compensate for the longer working days. People who work twelve consecutive hours as a result of overtime are not said to be working compressed work weeks because the increased length of one shift is not compensated for by the reduction in the number of shifts worked.

Research on performance over an extended working day has shown that people will experience more fatigue at the end of the shift as measured by a battery of fatigue-sensitive tasks (Rosa, 1991; Rosa and Bonnet, 1993), although other aspects of health and performance may show improvements on twelve-hour shifts (such as reported health and sleep problems; Williamson et al., 1994). Research on accidents and injuries has failed to find a clear pattern of either increased or decreased accidents and injuries on twelve-hour shifts. Northrup (1991) found that 9 of 18 Canadian and US steel mills surveyed in their study reported that their accident rates were no different on twelve-hour shifts compared with eight-hour shifts, seven reported a decrease in accidents after the shift change and two reported an increase. Northrup's findings extended the findings of an earlier survey of 50 petroleum and chemical companies in which no company reported an increase in accidents (Northrup et al., 1979). It should be noted that these conclusions were based on self-report data from companies taking part in the survey and, as such, the validity of the data has not been independently assessed. Laundry and Lees (1991) investigated the relationship between eight- and twelve-hour shifts by analyzing an organization's records 10 years before and after the change to twelve-hour shifts. They found minor accident rates to be lower on the twelve-hour shifts but off-the-job accident rates to be higher.

In a study of two manufacturing companies, Pollock et al. (1994) found no change in injury rate in the eighteen months before and after a change to twelve-hour shifts. In an extension of this study. Pollock and Smith (in press) gathered injury records from a further nine manufacturing and production sites/companies and found no significant change in injury rate in eight of them. One company did show a significant increase in injuries on twelve-hour shifts. However, five of the companies that had not shown a significant change in injury rate when all injuries were taken together did show a significant change in either severity or type of injury. The direction of the change in type and severity was not consistent across companies, indicating that the impact of twelve-hour shifts can vary according to the type of task, environment and work-force.

6.3 HUMAN ERRORS AND ACCIDENTS

The research on accident and injury rates on twelve-hour shifts is limited by the relative (and thankful) lack of data over sufficiently short time periods to allow comparisons of the effect of shift, without confounding the results by other changes in the workplace. Even when an approach is taken to analyze the types of accidents and injuries which occur, as was the case in Pollock *et al.* (1994) and Pollock and Smith (in press), the rarity of events makes it difficult to achieve adequate statistical power to avoid type II errors. For this reason, it is worth identifying significant precursors to accidents so that the incidence of these precursors can be used to assess the impact of work changes on the likelihood of accidents occurring. One type of event that fits these requirements is human error. Errors are more common occurrences than accidents but the underlying causes for many errors and accidents are similar. Thus, an analysis of the cause of human errors identifies factors that could result in accidents.

Human error is a major factor in accidents and injuries at work. Data from industries such as aviation suggest that somewhere in the region of 75–80% of accidents are caused by human errors of some type (Fegetter, 1982). A study of occupational fatalities in Australia between the years 1982 and 1984 revealed that 91% of the fatalities involved behavioural factors (Williamson and Feyer, 1990). These statistics should not be taken to suggest that human fallibility is the only contributing factor towards accidents and injuries at work; however, such statistics do illustrate the importance of human error for accidents. The Willamson and Feyer study emphasizes that it is the combination of behavioural and non-behavioural factors that causes fatalities at work. As Reason (1990) argues, accidents result from the coincidence of environmental, management and personal factors within a system that lacks adequate defences to prevent errors from resulting in bad outcomes.

If one takes a broad view of accident causation and considers not just operator errors but also errors in maintenance, design and management, then the impact of the human factor becomes a, if not the, most important part of the system that needs to be studied in order to reduce occupational injuries.

6.3.1 Error types

Errors do not always lead to accidents but the fact that errors have been made could mean that an accident is more likely. Errors may therefore be a pre-indicator for accidents and injuries, and their study could be used to identify areas where the potential for accidents is raised. An increase in errors may precede increased accident or injury rates and thus alert the researcher to areas where the workers are under increased demands. Furthermore, the *types* of errors that occur should indicate the types of performance that are under stress. One should therefore look for a model that maps human error to a model of human performance.

One such model is that proposed by Reason (1990). He extended Norman's (1981) slip/mistake distinction and combined it with Rasmussen's skill–rule–knowledge-based performance categories (Rasmussen and Jensen, 1974; Rasmussen, 1981), both of which were derived from studies of behaviour and human performance. Reason proposed that errors could be divided into three types:

1. *Skill-based errors* occur during highly practised skills where the level of control of performance is limited to attention checks on progress. Skill-based errors may be indicators of fatigue through attention failures. Their occurrence implies a need to improve a person's allocation of attention to appropriate tasks through reduction in stressors/distracters and redesign of the workplace.

2. *Rule-based mistakes* occur in situations where performance is mediated by production-type rules (if–then rules). Mistakes occur due to bad rules or the inappropriate use of good rules. Rule-based errors may be indicators of poor training or supervision and imply a need to reconsider the level and extent of training.

3. *Knowledge-based mistakes* occur when the person is required to perform beyond their normal (rule- and skill-based) knowledge due to an unusual event. The person then has to draw on their model of the work system and use active problem-solving to reach a solution. Mistakes in knowledge-based performance result from the in-built limitations of people's problem-solving capacity, as well as inaccuracies in a person's model of their work system. Knowledge-based errors may be indicators of insufficient operationalization of routine procedures or a lack of knowledge of emergency procedures. They imply that there is a need for training in some procedures, development of background knowledge about the workplace and operation, and/or improved availability of support when confronted by unfamiliar events.

If a person knowingly performs incorrectly, they have committed a violation. Violations may share some similarities with other error types but are characterized by the impact of motivational factors in their occurrence. Violations may be indicators of motivational problems and imply a need to adjust reinforcement of appropriate behaviour or remove conflicting rules that prevent a person from acting correctly.

Reason's model of human performance has been used widely to identify the errors that people make at work. For example, Edkins and Pollock (Edkins, 1995; Edkins and Pollock, 1996) studied the errors made in accidents and incidents by rail-workers over a four-year period. The reports generated by these incidents allowed the human errors that contributed to the incidents and accidents to be identified. The first point to note from this study was that of 112 incidents that could be attributed to human error, only five resulted in a collision. Clearly an analysis that only looked at the accidents would lack sufficient data from which to draw conclusions. The second point arising from the study is that the types of errors and factors contributing to these errors could be identified. Figure 6.1 shows the proportion of different error types identified from the study.

The study also showed that attentional factors (inattention or over-attention) accounted for 83% of skill-based errors. These findings were used to propose a

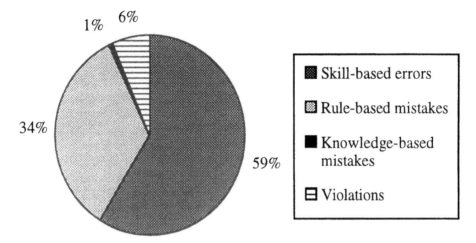

Figure 6.1 The relative proportion or different error types in an analysis of errors from a rail organization.

series of recommendations that focused on providing system defences against the most common skill-based error-missed signals. The study provided a rationale for implementing change before an accident occurred.

6.3.2 Twelve-hour shifts and error patterns

An analysis of error types could be especially useful where a change in working conditions brings with it both good and bad influences on performance, as is potentially the case with twelve-hour shiftwork. The study of errors on twelve-hour shifts could indicate where performance is more stressed and liable to break down and thus what remediation strategies could be used to increase safety.

One of the main concerns about the use of twelve-hour shifts is in the possibility of increased fatigue towards the end of a shift or block of shifts (Duchon and Smith, 1993). The discussion above would imply that this would lead to an increase in skill-based errors, in that fatigue would lead to inappropriate attentional focus and thus increase the chance of an error of execution.

Alternatively (or additionally), the increased blocks of time away from work may lead to a reduction in workers' familiarity with the conditions and constraints operating within a workplace. A lack of knowledge of work conditions could affect an individual's situational awareness (Endsley, 1995) or mental model (Norman, 1983). Situational awareness (SA) and mental models are terms often used interchangeably but they were developed to account for different concepts. The term mental model most often refers to a person's knowledge of a relatively stable environment, i.e. it might refer to a user's perception of how a machine works. Changes in a person's mental model reflect changes in their understanding of their environment. A person's mental model of their work environment might be inaccurate following a period away from the work environment if changes to the environment had been implemented without the worker's knowledge. Thus, if the worker had to perform at a knowledge-based level, their assumptions about the work environment would be based on their (perhaps faulty) mental model of the environment, thus leading to knowledge-based mistakes.

Situational awareness usually refers to a person's current understanding of a rapidly changing environment, such as in air-traffic control (ATC). In an ATC example, the controller's SA provides a picture (either as a visual image or as a set of aircraft position parameters; Hopkin, 1995) of the current locations of the aircraft under his or her control. It is known that this 'picture' is developed over a period of time when a person resumes work (in the region of 15 to 20 minutes for an air-traffic controller (Whitfield and Jackson, 1982). It is possible that the time required to develop the operator's picture would be lengthened following an extended break from work, as typically occurs with compressed work weeks. Poor situational awareness would be expected to result in the operator selecting inappropriate performance rules, thus leading to rule-based mistakes. (The error here would be of a rule- or knowledge-based type as the error is in the planning of the action, not in its execution.)

Preliminary evidence for the different effects of errors on twelve- and eight-hour shifts has been obtained from a pilot study conducted by the author with a large aircraft engineering company. The study investigated the relationship between working eight- versus twelve-hour shifts on self-reported worker performance, sleep patterns and shift preferences. A total of 47 people (25% of the twelve-hour shiftwork force) were interviewed in the study using a modified form of the standard shiftwork index (Barton *et al.*, 1990). All these people had worked for the company in the same or similar jobs since the move from eight- to twelve-hour shifts four years previously.

As has been found in other studies, the preference was strongly for maintaining the twelve-hour shifts (77% acceptance) although it was recognized that there were both advantages and disadvantages to the shift pattern. It was also recognized that some types of performance had decreased on the longer shifts. The workers' estimates of their sleep disturbance and mental fatigue were significantly higher for the twelve-hour shifts than for the eight-hour shifts (sleep disturbance: $t = 2.897, p < 0.01$; mental fatigue: $t = 3.032, p < 0.01$). The interviewees particularly mentioned 'minor lapses in attention towards the end of the shift' as a problem; in other words, skill-based errors. There had been no apparent change in accident rate with the change to twelve-hour shifts but the data suggested that there was potential for a problem to arise. The workers attributed the lack of increase in accidents as a result of people double-checking their work and the work of others. Some interviewees stated that their increased tendency to double-check had actually improved safety. On the other side of the argument was the workers' anecdotal observations that some aspects of their performance had improved because they were more motivated and had the chance to undertake and complete longer tasks.

It was notable that the more favourable responses to twelve-hour shifts came from the majority of engineers doing routine tasks. Those people who were in coordination or supervisory positions (who also worked twelve-hour shifts) were less likely to believe that the advantages outweighed the disadvantages. Whereas overall 77% preferred twelve-hour shifts, this dropped to just 45% among those in the top supervisory positions. Although there were no significant differences in any of the variables between the supervisors and others, the direction of the effects in most cases was for the supervisors to have experienced more problems on the twelve-hour shifts. There were no significant differences in responses for the older half of the population compared with the younger half.

It appears likely that the type of work undertaken will greatly influence the acceptability of twelve-hour shifts. Mentally demanding tasks, such as those undertaken by people in coordination positions, may be less suitable for twelve-hour shifts than those that may be less mentally demanding as the ability to process high volumes of information is greatly affected by fatigue (Holding, 1983).

The results indicated that, although there was no evidence of an overall change in safety as a result of the move to twelve-hour shifts, the workers believed that skill-based errors (slips and lapses) had increased. They also expressed awareness of their and their colleagues' potential increased likelihood to make minor mistakes and reportedly checked each other's work more carefully as a result. The workers also commented that any increase in skill-based errors was counteracted by their increased ability to continue tasks through to completion in the longer periods at work. Overall, it appeared that the workers in this company had experienced changes in performance that could have been identified through a close analysis of the types of errors that occurred, despite no apparent overall change in safety. Further data on this proposition is currently being sought.

6.4 PROBLEMS OF ERROR CATEGORIZATION

The approach outlined above requires the researcher to have access to errors that are committed by workers. The method used in the pilot study described above required workers to comment on the incidence of different error types. Although workers did not appear to have any difficulties in understanding the different types of errors, their awareness of their own errors is questionable. Some errors (notably skill-based slips) are easy to self-detect as they result in 'actions not as planned'. However, other errors (such as knowledge-based mistakes) may be difficult to identify unless the action (or inaction) results in some detrimental outcome to the individual or group. More realistic data may be gained from investigating records of events that have not resulted in accidents or injuries but were noted as 'incidents'. Currently only some industries have incident report procedures that would provide data suitable for error analysis.

When incident reports do exist, several studies have shown that they can be analyzed to identify the human errors that were made using a classification scheme such as that proposed by Reason (1990). Langan-Fox and Empson (1985) found that 94% of air-traffic control errors could be classified; Williamson and Feyer (1990) stated that only one in ten occupational fatality errors were uncodable. The Edkins and Pollock (1996) study described earlier found that only 10% of incidents contained insufficient information to be suitable for error analysis. However, classification of errors using this form of data must be undertaken with care. Incident reports are usually written for the purpose of describing the actions and the possible or actual consequences of those actions. Reports seldom require the writer to document their internal state or skill level. Thus the identification of errors and their classification can be subject to interpretation. In many situations the task and experience of the worker will identify whether the person was performing at a skill-, rule- or knowledge-based level. The researcher therefore needs to be knowledgeable of the tasks and workers to undertake this classification accurately. Inter- and intra-rater reliability is critical in what can be a subjective process and should be monitored through multiple coders and coding. Utilized carefully, human error data can provide a reliable and valid tool for accident prevention.

6.5 CONCLUSION

Despite the difficulties inherent in the analysis of errors from incident reports that were not designed for the purpose, human error data has the potential to provide a valuable indicator of the impact of work changes on safety and performance in advance of measurable increases in accidents and injuries. Such an analysis also provides the researcher and organization with a sensitive tool to observe changes in the types of problems that are being experienced, even when overall accident statistics are not affected. The problems of the lack of reporting of incidents and the lack of detailed background information on the incidents should be taken as a prompt to organizations to improve their incident reporting procedure so that such analysis can be undertaken with greater accuracy.

References

BARTON, J., FOLKARD, S., SMITH, L.R., SPELTON, E.R. and TOTTERDELL, P.A. (1990) *Standard shiftwork index manual*, SAPU Memo No. 1159, Social and Applied Psychology Unit, Department of Psychology, University of Sheffield.

DUCHON, J. and SMITH, T. (1993) Extended workdays and safety. *International Journal of Industrial Ergonomics*, **11**, 37–49.

EDKINS, G. (1995) Minimising the human factor in accident causation: a proactive vs reactive approach. In *Advances in industrial ergonomics and safety VII*, eds A.C. Butter and R. Steele, London: Taylor & Francis.

EDKINS, G. and POLLOCK, C.M. (1996) Proactive safety management: application and evaluation within a rail context. *Safety Science*, **24**, 83–93.

ENDSLEY, M.R. (1995) Toward a theory of situational awareness in dynamic systems. *Human Factors*, **37**(1), 32–64.

FEGETTER, A.J. (1982) A method for investigating human factor aspects of aircraft accidents and incidents. *Ergonomics*, **25**, 1065–1075.

HOLDING, D.H. (1983) Fatigue. In *Stress and fatigue in human performance*, ed. G.R.J. Hockey, Chichester: Wiley.

HOPKIN, V.D. (1995) *Human factors in air traffic control*, London: Taylor & Francis.

LANGAN-FOX, C.P. and EMPSON, J.A.C. (1985) 'Actions not as planned' in military air traffic control. *Ergonomics*, **28**, 1509–1521.

LAUNDRY, R. and LEES, B. (1991) Industrial accident experience of one company on 8- and 12-hour shift systems. *Journal of Occupational Medicine*, **33**(8), 903–906.

NORMAN, D.A. (1981) Categorisation of action slips. *Psychological Review*, **88**, 1–15.
  (1983) Some observations on mental models. In *Mental models*, eds D. Gentner and A.L. Stevens, Hillsdale, NJ: Lawrence Erlbaum Associates.

NORTHRUP, H.R. (1991) The twelve-hour shift in the North American mini-steel industry. *Journal of Labor Research*, **12**(3), 261–278.

NORTHRUP, H.R., WILSON, J.T. and ROSE, K.M. (1979) The twelve-hour shift in the petroleum and chemical industries. *Industrial and Labor Relations Review*, **32**(3), 312–326.

POLLOCK, C.M. and SMITH, G. (submitted) A study of the change in accident rate, type and severity in manufacturing and production companies.

POLLOCK, C., CROSS, R. and TAYLOR, P. (1994) Influences of 12- and 8-hour shifts on injury patterns. In *Proceedings of the 12th Triennial Congress of the International Ergonomics Association*, vol. 5, Toronto: Human Factors Association of Canada.

RASMUSSEN, J. (1981) Models of mental strategies in process plant diagnosis. In *Human detection and diagnosis of system failures*, eds J. Rasmussen and W. Rouse, New York: Plenum.

RASMUSSEN, J. and JENSON, A. (1974) Mental procedures in real-life tasks: a case study of electronic troubleshooting. *Ergonomics*, **17**, 293–307.

REASON, J. (1990) *Human error*. Cambridge: Cambridge University Press.

ROSA, R.R. (1991) Performance, alertness and sleep after 3.5 years on 12 h shifts: a follow-up study. *Work and Stress*, **5**(2), 107–116.

ROSA, R.R. and BONNET, M.H. (1993) Performance and alertness on 8 h and 12 h rotating shifts at a natural gas utility. *Ergonomics*, **36**(10), 1177–1193.

WHITFIELD, D. and JACKSON, A. (1982) The air traffic controller's 'picture' as an example of a mental model. In *Analysis, Design and Evaluation of Man. Machine Systems*, eds G. Johannsen and J.E. Rijnsdorp, New York: Pergamon.

WILLAMSON, A. and FEYER, A.M. (1990) Behavioural epidemiology as a tool for accident research. *Journal of Occupational Accidents*, **12**, 207–222.

WILLIAMSON, A.M., GOWER, C.G.I. and CLARKE, B.C. (1994) Changing the hours of shiftwork: a comparison of 8- and 12-hour shift rosters in a group of computer operators. *Ergonomics*, **37**(2), 287–298.

PART THREE

The role of risk in safety

INTRODUCTION

The concept of risk is a pivotal one in safety and is one of the cornerstones to our understanding of why safety problems occur. The concept uses the ideas of hazard, danger and of potential harm to workers or to the general integrity of the work system. One of the problems in using the concept, however, is that the conceptualization of risk is often limited and is often difficult to see all of its facets. Risk as a concept is extremely broad and involves hazards from every conceivable aspect of the work situation including:

- the hazards from the physical work environment, for example the equipment and machinery, the level of noise or temperature, exposure to harmful chemicals and the layout of the workplace;
- the hazards due to the behaviour of people at the workplace;
- the hazards due to the way the work is organized, including the work flow and pace, the hours that are worked and the communication flow within the workplace; and
- the hazards due to the whole system of work, mostly resulting from the interaction of each of any of the above aspects, for example hazards emanating from the way the work is organized to take account of particular physical hazards or from the need for workers to change their behaviour in work environments or situations that are hazardous.

A very large literature has grown up around the need to account for risk in a systematic way; to predict the form it will take, when and where it will occur and, where its consequences are likely to be harmful, to modify it. The emphasis has been on attempting to estimate the probability that work systems will fail or be reliable. Over the last 20 years there has been considerable development of a wide range of probabilistic risk assessment tools such as event tree analysis, fault tree analysis, the hazard and operability study technique (HAZOP) and the management oversight risk tree technique (MORT) (Kirwan and Ainsworth, 1992). While these techniques helped to make great strides in identifying, quantifying and evaluating the expected frequency of unsafe occurrences, they were weak in accounting for the role of human behaviour, particularly error. The more recent addition of human reliability analysis (HRA) techniques has increased the usefulness of reliability estimation methods by taking fuller account of the role of human behaviour in causing unsafe occurrences (Kirwan, 1994).

These adaptations of reliability analysis are really only a starting point. Even with the development of HRA, the usefulness of these techniques in actually predicting the likelihood of unsafe occurrences is limited, especially if the occurrences include error. Rasmussen, for example, argues that risk analysis, and in particular the identification of hazards, is at present more of an art than a science (Rasmussen, 1993). In recent years the search to understand human risk has been through trying to learn more about the causes of errors, including the influence of factors like the social and organizational settings in which they occur (e.g. Wagenaar and Groeneweg, 1987; Reason, 1990). This has led to a number of studies of the relationships between the safety climate in a workplace, worker risk perception and worker behaviour. The importance of taking this tack in the study of risk is emphasized by the findings from a number of studies that the roots of human reliability, or any compromise to safety, may be inherent in the work system itself (Wagenaar and Groeneweg, 1987; Feyer and Williamson, 1991). Thus, the safety culture of an organization can constitute an important determinant of human risk. The chapter by Dedobbeleer and Beland in this part reviews the evidence for a relationship between safety climate and risk perception. They conclude that organizational context influences workers' perception of risk and that attempts to reduce human risk should, as a consequence, focus on the organizational context in which risks are taken or avoided.

The chapter in this part by Wilde paints a somewhat different view of the role of risk perception and its implications for solving the problems due to unsafe or risky behaviour. The conception of risk in this chapter is based on the theory of risk homeostasis. This follows the argument that individuals optimize their risk in any situation according to their overall needs for a particular outcome and not according to other factors like the physical environment or their own level of skill. Taking this view of risk leads to considerably different views of the how best to reduce risk. In contrast to the solutions advanced by Dedobbeleer and Beland, Wilde argues that motivators and incentives for individual workers will provide the best chance for reducing human risk.

This notion is also taken up in the chapter by DeJoy, Gershon and Murphy in this part. Using the example of the need to reduce the risk of HIV/AIDS for healthcare workers, they explore the relationships between three aspects: job and task factors, worker factors and organizational/environmental factors. Their review also emphasizes the importance of performance feedback systems, but more significantly it points to the need for a work systems approach to risk minimization. They argue that developing risk reduction strategies that focus only on one or two aspects may necessarily be inadequate especially where there are barriers or impediments in the system which make unsafe behaviour more advantageous than safe behaviour.

The three chapters in this part tackle the issue of risk from quite different perspectives. Taken together they also raise a number of important questions. These will need to be addressed before the concept of risk can be developed fully and its role in attempts to reduce occupational injury can be properly understood.

References

FEYER, A.-M. and WILLIAMSON, A.M. (1991) A classification system for causes of occupational accidents for use in preventive strategies. *Scandinavian Journal of Work, Environment and Health*, **17**, 302–311.

KIRWAN, B. (1994) *A guide to practical human reliability assessment*. London: Taylor & Francis.

KIRWAN, B. and AINSWORTH, L.K. (1992) *A guide to task analysis*. London: Taylor & Francis.

RASMUSSEN, J. (1993) Learning from experience? How? Some research issues in industrial risk management. In *Reliability and safety in hazardous work systems*, eds B. Wilpert and T. Qvale, pp. 43–65, Hove: Lawrence Erlbaum.

REASON, J. (1990) *Human error*. Cambridge: Cambridge, University Press.

WAGENAAR, W.A. and GROENEWEG, J. (1987) Accidents at sea: multiple causes and impossible consequences. *International Journal of Man Machine Studies*, **27**, 587–598.

CHAPTER SEVEN

Is risk perception one of the dimensions of safety climate?

NICOLE DEDOBBELEER AND FRANÇOIS BÉLAND

7.1 INTRODUCTION

Safety climate has become a popular concept. It has been given all kinds of positive attributes such as helping to understand employee performance, assuring success in injury control and providing guidelines in the establishment of a safety policy (Brown and Holmes, 1986; Mattila *et al.*, 1994a). However, there is no specific definition of the concept and as such no clear guidelines for measurement of the concept exist.

A few studies have measured the safety climate concept (Zohar, 1980; Brown and Holmes, 1986; Lutness, 1987; Dedobbeleer and Béland, 1991; Oliver *et al.*, 1993; Melia *et al.*, 1992; Mattila *et al.*, 1994a, 1996; Coyle *et al.*, 1995; de Jong and Landweerd, 1991). The results show that there is a great diversity in the type and number of safety climate dimensions presented in the validation studies. The concept of risk is one of the safety climate dimensions in some of these studies. The question is: is it one of the dimensions of safety climate? The aim of this chapter is to review published studies on safety climate and to discuss the importance of the concept of risk in safety climate measurement.

7.2 CONCEPTS OF SAFETY CLIMATE AND RISK

7.2.1 Concept of safety climate

The safety climate concept has been developed in the context of the generally accepted definitions of the organizational climate. Organizational climate has been used to refer to a broad class of organizational and perceptual variables that reflect individual–organizational interactions (Jones and James, 1979; Glick, 1985). James and Jones (1974) proposed a differentiation between the organizational climate and the psychological climate. The former is examined at the organizational level of analysis and the latter is studied at the individual level of analysis. Both are seen as multidimensional phenomena, descriptive of employees' perceptions of their experiences within a work organization.

In the different studies which measured the safety climate concept, it was viewed as a summary of molar perceptions workers share about their work environment. According to Koys and DeCotiis (1991), climate perceptions summarize an individual's description of his or her organizational experiences rather than his or her affective evaluative reaction to what has been experienced. Based on a variety of cues present in their work settings, employees are believed to develop coherent sets of perceptions and expectations regarding behavior–outcome contingencies and to behave accordingly (Fredericksen *et al.*, 1972; Schneider, 1975a,b).

Over the years there has been an increasing concern with a broader environmental parameter: the organization's social climate surrounding health issues. This concept has been given different labels: health strengthening environment (Pender, 1989), health-conscious work environment (Everly and Feldman, 1985), healthy organizational culture (Opatz, 1985), social climate encouraging health improvement (Heirich *et al.*, 1989), climate for health (Ilgen, 1991), a social climate with a pro-health philosophy (Abrams *et al.*, 1986) and a wellness-oriented workplace (Chapman, 1987). The relationship between this broader environmental parameter and the safety climate concept has not yet been addressed.

7.2.2 Concept of risk

The term 'risk' can be applied to almost any activity we engage in. A great deal of emphasis has thus been placed on risk. There is, however, much disagreement about the definition of the concept of risk as well as about its importance. It has been defined statistically to provide a relative measure of safety (accident rates). It is also used as synonym for danger or threat, stressing the more subjective aspects of risk (Wilde, 1982, 1986; Fuller, 1984). According to Haight (1986), it can be used as probability and expectation. Risk is further defined by Oppe (1988) following the decision–theory approach as expected loss of an alternative to be chosen. In this sense it concentrates attention on the process of decision-making and not solely on the result.

Connors (1992) indicates that in recent years anthropology has opened up the discussion of risk by emphasizing the role of context and culture in shaping both the perception and the experience of risk (Douglas, 1985; Gifford, 1986; Rappaport, 1988). According to Gifford (1986), 'risk perception is not a unified phenomenon but one that is conditional on social status, social rules and rewards in particular contexts'.

In the organizational climate field, workers' perception and experience of risk may thus be seen as influenced by the context and culture of their work environment. If this is true, is employee risk perception part of the workplace safety climate?

7.3 RISK AS A SAFETY CLIMATE DIMENSION

In the general organizational climate literature there is very little agreement on the dimensions of organizational climate. Table 7.1 shows the diversity in the number and types of safety climate dimensions presented in validation studies. Only two studies (Zohar, 1980; Brown and Holmes, 1986) found that employee physical risk perception was one of the safety climate dimensions. In attempting to resolve the differences in reported factor patterns, several issues can be raised: item selection, statistical methods and context (i.e. type of industry, organizational characteristics, organizational culture, characteristics of the work situation, socioeconomic characteristics, country, etc.).

Researchers are encouraged to use climate dimensions that are likely to influence or be associated with the study's criteria of interest. Results show that Zohar's study was a source of inspiration for the choice of safety climate items in most of the studies on safety climate. However, an adaptation of these items was made in almost all studies. A majority of the researchers included items that specifically addressed health and safety issues perceived to be important within the various organizations studied. Variables were then often similar but not identical.

Zohar (1980) developed a questionnaire from seven sets of items that were descriptive of organizational events, practices and procedures which were found to discriminate high accident factories from low accident factories (Cohen, 1977). He sampled 20 plants in separate firms of chemical, metal and food industries in Israel. Table 7.1 shows that eight climate dimensions were found and that risks at the workplace is one of them. Brown and Holmes (1986) used Zohar's 40-item questionnaire in an American sample of production workers to replicate Zohar's factor structure. They found three dimensions and one of them was employee physical risk perception. Dedobbeleer and Béland (1991) assessed the validity of Brown and Holmes' three-factor model on an American sample of construction workers. They used nine variables representing safety concerns in the construction industry. Variables were similar but not all identical to those included in Brown and Holmes' three-factor model. The best-fit model was a two-factor model (Figure 7.1) and it did not include employee physical risk perception as one of the dimensions. A replication of Dedobbeleer and Béland's two-factor model was made by Oliver et al. (1992) and Meliá et al. (1993) with nine similar but not identical variables measuring climate perceptions among post-traumatic and pre-traumatic workers from different types of industries in Spain.

Coyle et al. (1995) included items considered generalizable from Zohar's (1980) and Glennon's (1982) questionnaires as well as items identified through interviews with a representative cross-sample of personnel. He ranked them in sessions where a nominal group technique was used. Twenty-six of the 30 items were identical for the two organizations. These items were administered to two similar organizations involved in the provision of health care and social services to the elderly in Australia. The results show that seven dimensions were retained in the first organization and three in the second one. Maintenance and management issues explained 60% of the variance in the first organization. Work environment explained 71% of the variance in the second organization. None of the factor patterns were similar to Zohar's results. Employee risk perception was not one of the dimensions identified.

Three studies found in the literature review (Lutness, 1987; de Jong and Landerweerd, 1991; Mattila et al., 1994a) were not validation studies and did not provide information on the method by which safety climate factors were identified. Lutness (1987) presents a climate survey developed by Du Pont's Safety Services. One of the ten dimensions, 'general safety perceptions', is related to employee physical risk perception. de Jong and Landerweerd (1991) presented three dimensions but employee risk perception is not one of the dimensions identified.

Table 7.1 Studies conducted on safety climate

Author Country	Type of industry	Sample size	Source of items	Statistical methods	Dimensions
Zohar (1980) Israel	Various	20 industries 400 production workers	Cohen's study	Principal component factor analysis with varimax rotation	Perceived management attitudes on safety Effect of safe work practice on promotion Effect of safe conduct on social status Status of safety committee Status of safety officer Importance of effectiveness of safety training Risks at the workplace Effect of required work pace on safety
Brown and Holmes (1988) USA	Manufacturing	10 companies 425 workers	Zohar's questions	LISREL maximum likelihood method	Employee perception of how concerned management was with their well-being Employee perception of how active management was in responding to this concern Employee physical risk perception
Lutness (1987) USA	Chemistry	NA	NA	NA	Management issues Supervision issues Co-worker safety General safety perceptions Safety practices Employee health Safety program effectiveness Degree of employee involvement Working conditions
de Jong and Landeweerd (1991) Holland	Construction	49 companies 98 sites 663 workers 98 job-site managers	NA	NA	Companies, top management and workers' concern with safety Relative concern with safety of the company Job-site manager concern for safety

Study	Industry	Sample	Questionnaire	Analysis method	Factors
Dedobbeleer and Béland (1991) Canada	Construction	9 sites 384 workers 9 foremen	Adaptation of Brown and Holmes' questions	LISREL maximum likelihood method, weighted least squares method	Management commitment Workers' involvement
Meliá et al. (1991) Spain	Various	243 workers	Adaptation of Dedobbeleer and Béland's questions	LISREL maximum likelihood method, weighted least squares method	Management commitment Workers' involvement
Oliver el al. (1993) Spain	Various	182 workers	Adaptation of Dedobbeleer and Béland's questions	LISREL maximum likelihood method, weighted least squares method	Management commitment Workers' involvement
Mattila et al. (1994) Finland	Construction	16 sites 283 workers 15 managers 16 foremen	NA	NA	NA
Coyle et al. (1995) Australia	Health care and social services	2 organizations 340 employees	Interview, nominal group technique and Zohar's questions	Principal component factor analysis with varimax rotation	*Company 1:* Maintenance and management issues Company policy Accountability Training and management attitudes Work environment Policy/procedures Personal authority *Company 2:* Work environment Personal authority Training and enforcement of policy

NA: not available

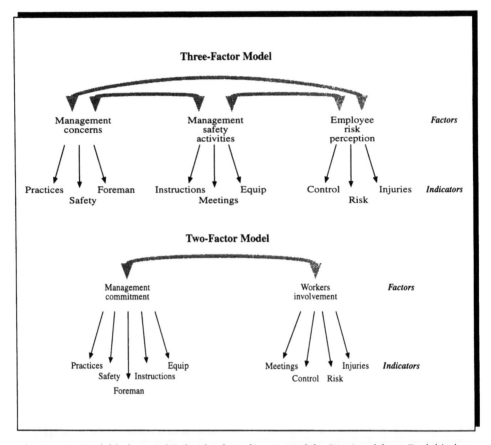

Figure 7.1   Dedobbeleer and Béland safety climate models. Reprinted from Dedobbeleer, N. and Béland, F. copyright (1991) A safety climate measure for construction sites. *Journal of Safety Research*, vol. 22, pp. 97–103, with kind permission from Elsevier Science Ltd, The Boulevard, Langford Lane, Kidlington OX5 1GB, UK.

Mattila *et al.* (1994a) did not identify the dimensions of the safety climate measure. The results of the literature review thus indicate that it remains likely that item selection plays a certain part in affecting the safety climate factor structure.

It might be argued that the different results obtained by the researchers reflect differences in the statistical methods used. Brown and Holmes (1986) did not use Zohar's statistical technique (i.e. principal component factor analysis with varimax rotation). Instead they used the maximum likelihood procedure in LISREL. Dedobbeleer and Béland (1991) used nine variables to assess the validity of Brown and Holmes' three-factor model on an American sample of construction workers. When they used the same statistical technique as Brown and Holmes, this model was retained. However, when they used a more appropriate procedure adapted to the polychotomous ordinal level data of the study, the weighted least squares method in LISREL, to estimate the parameters of the model, a two-factor model was found. Employee physical risk perception was not one of the two dimensions. Oliver *et al.* (1993) and Meliá *et al.* (1992) replicated the Dedobbeleer and Béland results by using similar statistical techniques. Finally, Coyle *et al.* (1995) used identical statistical techniques to Zohar (although the cut–off point for factor loading was set at 0.40 instead of 0.49) but selected items perceived to be important within the two organizations included in his study. Although the majority of these items and the statistical techniques (i.e. principal component factor analysis with varimax rotation) were identical in the two organizations, no similar set of factors was identified with these two highly similar organizations. It is thus fair to say that methodological issues alone or item selection issues alone do not account for the different findings.

It might also be argued that differences in the factor structure are related to differences in context (e.g. types of industries, countries, etc.). Dedobbeleer and Béland showed, however, some stability of safety climate factors. A systematic replication of their bifactorial model was made with the same statistical method in a different country and in two different samples of workers in different occupations (Oliver *et al.*, 1993; Meliá *et al.*, 1992). The two-factor model provided the best fit. The climate structures did not differ between American construction workers and Spanish workers from different types of industries, subsequently providing a valid climate measure across different populations and different types of occupations. In this bifactorial model perceived likelihood of injuries and perception of risk-taking at the job are two of the three indicators of the dimension labelled 'workers' involvement'.

7.4 DISCUSSION

The literature review shows that there is not yet a universal set of safety climate factors. Nevertheless, some of the research results are encouraging. A valid two-factor model, the Dedobbeleer and Béland model (1991), has been provided across different populations and occupations. The question that then arises is: is this set of factors generalizable across cultures and organizations? Further studies should be conducted if the replication and generalization rules of theory testing are to be met. The challenge is to identify a theoretically meaningful and analytically practical universe of possible climate dimensions.

Employee risk perception was identified in two of the nine studies examined. In other studies, worker risk perception is associated with worker perception of control. This association led to speculation that these two variables are highly related to workers' involvement or responsibility for safety. These results are related to the literature on democratic management, which stresses the importance of creating structures and processes that provide access to decision-making and enable participants to actively influence organizational decisions (Sass, 1989). In this context, safety climate dimensions are closely related to a 'partnership mentality' to improve job safety, contrasting with a police enforcement mentality. Sass (1989) argues for 'the development of an ethics of the work environment based upon egalitarian principles, and the transformation of the primary work group into a community of workers who can shape the character of their work environment'. This progressive philosophy brings us back to Glick's plead (1985) to define organizational climate as a generic term for a broad class of organizational, rather than psychological, variables that describe the organizational context for individuals' actions. As this organizational context with its norms and values influences workers' perception of risk and decisions, the effort to increase job safety needs to go beyond risk behaviour and risk perception. Attention should be given to the context in which individuals take risks and negotiate to avoid risk.

References

ABRAMS, D.B., ELDER, J.P., CARLETON, R.A., LASATER, T.M. and ARTZ, L.M. (1986) Social learning principles for organizational health promotion: an integrated approach. In *Health and industry: a behavioral medicine perspective*, eds M.F. Cataldo and T.J. Coates, pp. 28–51, New York: John Wiley & Sons.

BROWN, R.L. and HOLMES, H. (1986) The use of a factor-analytic procedure for assessing the validity of an employee safety climate model. *Accident Analysis and Prevention*, **18**(6), 445–470.

CHAPMAN, L. (1987) *Creating a wellness oriented workplace: policies, places and norms*, vol. 18, Seattle: Corporate Health Designs.

COHEN, A. (1977) Factors in successful occupational safety programs. *Journal of Safety Research*, **9**, 168–178.

CONNORS, M.M. (1992) Risk perception, risk taking and risk management among intravenous drug users: implications for AIDS prevention. *Social Science Medicine*, **34**(6), 591–601.

COYLE, I.R., SLEEMAN, S.D. and ADAMS, N. (1995) Safety climate. *Journal of Safety Research*, **26**(4), 247–254.

DEDOBBELEER, N. and BÉLAND, F. (1991) A safety climate measure for construction sites. *Journal of Safety Research*, **22**, 97–103.

DEDOBBELEER, N., BÉLAND, F. and GERMAN, P. (1990) Is there a relationship between attributes of construction sites and workers' safety practices and climate perceptions? In *Advances in industrial ergonomics and safety II*, ed. B. Das, pp. 725–732, London: Taylor & Francis.

DOUGLAS, M. (1985) *Risk acceptability according to the social sciences*, New York: Russell Sage.

EVERLY, G.S. and FELDMAN, R.H. (1985) *Occupational health promotion: health behavior in the workplace*, New York: John Wiley & Sons.

FREDERIKSEN, N., JENSEN, O. and BEATON, A.E. (1972) *Prediction of organizational behavior*, Elmsford, NY: Pergamon.

FULLER, R. (1984) A conceptualization of driving behavior as threat avoidance. *Ergonomics*, **27**, 1139–1155.

GIFFORD, S. (1986) The meaning of lumps: a case study of the ambiguities of risk. In *Anthropology and epidemiology*, ed. C.R. Janes *et al.*, pp. 213–246, Dordrecht: Reidel.

GLENNON, D.E. (1982) Safety climate in organizations. Ergonomics and occupational health. *Proceedings of the 19th Annual Conference of the Ergonomics Society of Australia and New Zealand*, 17–31.

GLICK, W.H. (1985) Conceptualizing and measuring organizational and psychological climate: pitfalls in multi-level research. *Academy of Management Review*, **10**(3), 601–616.

HAIGHT, F.A. (1986) Risk, especially risk of traffic accident. *Accident Analysis and Prevention*, **18**, 359–366.

HEIRICH, M.A., CAMERON, V., ERFURT, J.C., FOOTE, A. and GREGG, W. (1989) Establishing communication networks for health promotion in industrial settings. *American Journal of Health Promotion*, **4**, 108–117.

ILGEN, D.R. (1991) Health issues at work: opportunities for industrial/organizational psychology. *American Journal of Psychology*, **45**, 273–283.

JAMES, L.R. and JONES, A.P. (1974) Organizational climate: a review of theory and research. *Psychological Bulletin*, **81**(12), 1096–1112.

JONES, A.P. and JAMES, L.R. (1979) Psychological climate: dimensions and relationships of individual and aggregated work environment perceptions. *Organizational Behavior and Human Performance*, **23**, 201–250.

DE JONG, A.H.J. and LANDERWEERD, J.A. (unpublished) Safety in Construction: where to begin? An exploratory comparative organizational analysis of construction companies and construction sites.

KOYS, D.J. and DECOTIIS, T.A. (1991) Inductive measures of psychological climate. *Human Relations*, **44**(3), 265–285.

LUTNESS, J. (1987) Measuring up: assessing safety with climate surveys. *Occupational Health and Safety*, **56**(2), 20–26.

MATTILA, M., RANTANEN, E. and HYTTINEN, M. (1994a) The quality of work environment, supervision and safety in building construction. *Safety Science*, **17**, 257–268.

MATTILA, M., HYTTINEN, M. and RANTANEN, E. (1994b) Effective supervision and safety at the building site. *International Journal of Industrial Ergonomics*, **13**, 85–93.

MELIÁ, J.L., TOMAS, J.M. and OLIVER, A. (1992) Concepciones del clima organizacional hacia la seguridad laboral: replicación del modelo confirmatorio de Dedobbeleer y Béland. *Acceptado en Revista de Psicologia del Trabajo y de las Organizaciones*, **9**(22).

OLIVER, A., TOMAS, J.M. and MELIA, J.L. (1993) Una segunda validación cruzada de la escala de clima organizacional de seguridad de Dedobbleer y Béland. Ajuste confirmatorio de los modelos unifactorial, bifactorial y trifactorial. *Psicologica*, **14**, 59–73.

OPATZ, J.P. (1985) *A primer of health promotion: creating healthy organizational cultures*, Washington, DC: Oryn Publications.

OPPE, S. (1988) The concept of risk: a decision-theoretic approach. *Ergonomics*, **31**(4), 435–440.

PENDER, N.J. (1989) Health promotion in the workplace: suggested directions for research. *American Journal of Health Promotion*, **3**(3), 38–43.

RAPPAPORT, R. (1988) Towards a postmodern risk analysis. *Risk Analysis*, **8**, 2.

RIBISI, K.M. and REISCHL, T.M. (1993) Measuring the climate for health at organizations–development of the worksite health climate scales. *Journal of Occupational Medicine*, **35**(8).

SASS, R. (1989) The implications of work organization for occupational health policy: the case of Canada. *International Journal of Health Services*, **19**(1), 157–173.

SCHNEIDER, B. (1975a) Organizational climates: an essay. *Personnel Psychology*, **28**, 447–479.

(1975b) Organizational climate: individual preferences and organizational realities revisited. *Journal of Applied Psychology*, **60**, 459–465.

SCHNEIDER, B. and REICHERS, A.E. (1983) On the etiology of climates. *Personnel Psychology*, **36**, 19–39.

WILDE, G.J.S. (1982) The theory of homeostasis: implications for safety and health. *Risk Analysis*, **2**, 249–258.

(1986) Beyond the concept of risk homeostasis: suggestions for research and application towards the prevention of accidents and lifestyle-related disease. *Accident Analysis and Prevention*, **18**(5), 377–401.

ZOHAR, D. (1980) Safety climate in industrial organizations: theoretical and applied implications. *Journal of Applied Psychology*, **65**(1), 96–102.

CHAPTER EIGHT

The concept of target risk and its implications for accident prevention strategies

GERALD J.S. WILDE

8.1 OVERVIEW

Among the various possible approaches to accident prevention, the use of incentive programmes for accident-free performance has become more prominent in recent years. An incentive program entails the extension of a pre-announced gratification or bonus to potential recipients – like workers or drivers – on the specific condition that they do not have an accident caused by their own fault within a specified period of time. Thus, incentives must be distinguished from unexpected rewards. Incentive programmes differ from safety engineering and safety education by attempting to strengthen the *motivation* to be safe. Safety engineering and ergonomics may offer an *opportunity* for greater safety and education, and a greater *ability* to be safe (and lead to a decrease in the accident rate per unit distance travelled or unit of productivity), but greater opportunity or greater ability are not likely to enhance actual safety performance per person-hour of performance unless there is a desire for increased safety.

That the desire for safety can be increased by the use of incentives is explained in the context of risk homeostasis theory. Incentives serve to reduce the target level of risk, that is, the level of accident risk the operator is willing to accept in return for the benefits expected from the activity in question. Four utility factors determine the target level of risk. Target risk will be higher to the extent that:

1. the expected *advantages* of comparatively *risky* behaviour are higher;
2. the expected *costs* of comparatively *risky* behaviour alternatives are lower;
3. the expected *benefits* of comparatively *safe* behaviour alternatives are lower; and
4. the expected *costs* of comparatively *safe* behaviour alternatives are higher.

Some of the motivating factors in all four categories are economic in nature; others are cultural, social or psychological. They are usually so thoroughly internalized that most people, most of the time, are not consciously aware of them. Thus, the target level of risk should not be viewed as something that people arrive at by explicitly calculating probabilities of various possible behaviour outcomes and their respective positive or negative values, nor does the notion of target risk imply that people take risks for the sake of risk.

Attempts to motivate individuals to lower their target risk may, in principle, be carried out in a number of different ways. One of these takes the form of punishing operators for accidents in which they are deemed to be at fault (which increases the value of point 2 above); in another they are rewarded for the fact of not having an accident (which increases the value of point 3 above). The practice of punishing operators for specific unsafe actions has been found unsatisfactory for two reasons: it commonly fails to bring about a reduction in the accident rate and provokes undesirable side-effects including resentment and antagonism. The incentive approach, however, has been shown to be effective towards the objective in every case of implementation that has been documented to date in the professional journals. Incentives may in some cases lead to under-reporting of accidents, but this is their only identified undesirable side-effect. The under-reporting is limited to minor accidents, while a frequently observed beneficial side-effect is better company morale and productivity.

The effectiveness of incentive programs in reducing accidents is often remarkably high. Accident reductions by 80% or even more have been reported. The implementation costs of these programmes – that is the costs of the bonuses and programme administration – are considerably smaller than the savings due to accident reduction. Benefit/cost ratios are usually greater than two to one, meaning that companies can make money on these accident reduction efforts. This is largely due to the improvement in safety leading to reduced fees to workers' compensation and other forms of insurance. The favourable benefit/cost ratios imply that safety incentive programs can make a significant contribution to a company's financial well-being.

Some incentive programmes have been found to be more effective than others. It is, therefore, important to identify the specific design features of incentive programmes that enhance their effectiveness. In this chapter, some 15 factors have been specified that increase program effectiveness according to the present state of knowledge. These factors are presented as a 'checklist' for the purpose of maximizing the effectiveness of incentive programmes for accident-free operation. It is concluded that safety incentives have established themselves as the most effective form of safety management in industry and transportation at a high benefit/cost ratio.

8.2 RISK ACCEPTANCE AND HOMEOSTASIS

As there is no behaviour with total certainty of the desired outcome, essentially all behaviour may be viewed as risk-taking behaviour. 'Zero risk' is not a meaningful goal, as it can only exist in the absence of behaviour. Instead of aiming at the elimination of risk, an individual should attempt to optimize the exposure to risk in an activity, where 'optimal' means the degree of risk at which the aggregate needs of that individual are likely to be best fulfilled.

Why people should opt for a level of accident risk that is greater than zero can be explained by referring to Figure 8.1. Increasing your driving speed and/or the amount of driving you do corresponds to moving from left to right along the horizontal axis of exposure to accident risk. With increased exposure, both expected gains and expected losses increase. Greater speed means shorter travel time towards your destination as well as more thrill and excitement. A greater distance driven means more mobility. Greater speed, however, also means more wear and tear on your car, higher gasoline consumption, a chance of a traffic ticket and more severe consequences if an accident were to happen.

For each combination of speed and amount of driving, the expected net benefit equals the expected gain minus the expected loss. In Figure 8.1 the curves describing expected gain and expected loss have been drawn such that the expected net benefit curve rises from left to right, then reaches a maximum that is followed by a decline. At zero exposure to risk, there is no mobility and the net benefit of mobility is nil. When speed is extremely high, the expected loss is greater than the expected gain and the expected net benefit falls below zero.

The extremes are thus to be avoided: while it is obvious that people should not maximize the danger of an accident, neither should they opt for the other extreme of minimizing accident risk. What they should do instead is attempt to maximize the expected net benefit from road travel and choose a speed and other actions accord-

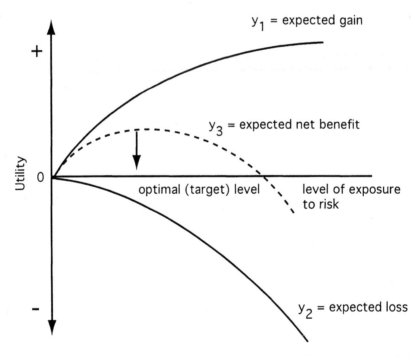

Figure 8.1 Theoretical representation of road users as net benefit maximizers and thus as risk optimizers. They choose an amount and manner of mobility such that the associated level of subjective risk corresponds with the point at which the expected net benefit is maximal. Note that the curve y_3 has been drawn so that each y_3 value equals the corresponding y_1 value minus the corresponding y_2 absolute (from Wilde, 1988a, 1994).

ingly. They should, therefore, try to select a level of exposure to accident risk that is greater than zero and promises maximal net benefit from the behaviours chosen, that is the target level of risk. Since zero risk is obviously not a meaningful goal, people target their risk level above zero. This is what they do when driving, but also at work, in sports and leisure-time activities. This is also what they do with respect to lifestyle choices that may have consequences for health, disease and death (such as tobacco and alcohol use, dental and sexual hygiene, diet and exercise). At least, this is what the theory of risk homeostasis (RHT) contends. When applied to the area of transportation, this theory posits that a jurisdiction's total road accident loss per time unit of aggregate road-user exposure is the output of a closed-loop control process in which the target level of risk (i.e. the inverse of the level of safety desired) operates as the unique controlling (reference, 'set-point') variable. Thus, apart from temporary fluctuations, risk per time unit of exposure to traffic is independent of factors such as the physical features of the traffic environment and road-user skills, and ultimately depends on the amount of accident loss accepted by the aggregate of road users in exchange for the benefits accruing from the desired amount and manner (e.g. speed) of mobility.

The term 'homeostasis' denotes a dynamic process, not a fixed value of some outcome variable. When applied to human behaviour, it also entails a learning component including error detection and adaptation to new conditions by trying out novel behaviour alternatives on their consequences. This may be clarified with the help of Figure 8.2. In any ongoing activity, people continuously check the amount of risk they feel they are exposed to (box b). They compare (comparator, summing point) their subjectively experienced risk with the amount of risk they are willing to accept (box a) and try to reduce and difference between the two to zero (box c). Thus, if the level of subjectively experienced risk is lower than is acceptable, people tend to engage in actions that increase their exposure to risk. If, however, the level of subjectively experienced risk is higher than is acceptable, they make an attempt

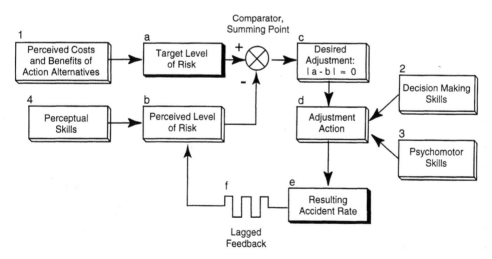

Figure 8.2 Homeostatic model relating the accident rate per head of population in a jurisdiction to the level of caution in road-user behaviour and vice versa, with the average target level of risk as the controlling variable (from Wilde, 1994).

to exercise greater caution. Consequently, they will choose their next action (box d) so that the subjectively expected amount of risk associated with that action matches the level of risk accepted. During that next action, perceived and accepted risk are again compared and the subsequent adjustment action is chosen in order to minimize the difference, and so on.

Each particular adjustment action carries an objective probability of risk of accident or illness. Thus, the sum total of all adjustment actions across all members of the population over an extended period of time (say one year) determines the temporal rate of accidents (box e) or lifestyle-dependent disease in the population.

These rates, as well as more direct and frequent personal experiences of danger, in turn influence (symbol f) the amount of risk people expect to be associated with various activities (box b again), and with particular actions in these activities, over the *next* period of time. They will decide on their future actions accordingly and these actions will produce the subsequent rate of human-made mishaps (box e again). Thus, a 'closed loop' is formed between past and present, and between the present and the future. In the long run, the human-made mishap rate essentially depends on the amount of risk people are willing to accept and will show no lasting change unless the target level of risk is altered. Factors outside the closed loop can only have modest or temporary influences on the per capita accident rate.

It is interesting to note the *circular* causality in this process: not only does behaviour determine the accident rate but changes in the accident rate trigger changes in behaviour. This mutual dependency may be illustrated by what happened in Sweden when, in the fall of 1967, a change-over from left-hand to right-hand traffic took place. Contrary to the expectation that the accident rate would increase as a result of inadequate perceptual and decision-making skills in the new situation, the traffic fatality rate per head of population actually diminished considerably. Between one and two years after the change-over, however, fatalities returned to the prior trend. The disturbance in the death rate was attributed to the sudden surge in the level of perceived risk due to the change-over (Alexandersson, 1972). The level of risk perceived exceeded the target level and people became very cautious with the result that the accident rate dropped. After some time (symbol f), people discovered that the roads were not as dangerous as they thought and the perceived level of risk came closer once again to the target level, leading people to become less prudent, and the accident rate rose again to 'normal'.

8.3 MOTIVATING PEOPLE TO ALTER THEIR TARGET LEVEL OF RISK

In the last few years there has been a major increase in interest in the use of incentives for accident-free operation as an approach to accident prevention. This appears to be an international phenomenon; this author has seen reports on the topic from Europe in Dutch, English, German, Norwegian, Russian, Spanish and Swedish, as well as the Americas. Publications on the topic have become more frequent not only in the scientific journals but also in the trade press (e.g. Markus, 1991; Legler, 1992; Mazzurco, 1992; Synnett, 1992; Hislop, 1993; Redmann, 1993; McIlwaine, 1994; Gerson, 1994; Weinstock, 1994; Colledge, 1995).

What explains this increased interest? On the one hand, it may be due to the large amount of empirical evidence compiled in occupational, clinical and health psychology that shows the fruitfulness of the behaviour modification approach to the treatment of dysfunctional behaviour and the shaping of desirable behaviours. On the other hand, there are indications of growing disappointment with the traditional approaches to accident prevention.

8.3.1 'Triple E'

Behaviour modification, with its focus on motivation, is indeed quite different from the traditional 'Triple E' approach towards road and occupational safety, i.e. engineering, education and enforcement. Engineering can provide an improved *opportunity* to be safe, education can enhance the performance *skills* and enforcement of rules against *specific unsafe acts* may be able to discourage people from engaging in these particular acts. However, none of these interventions necessarily increases the *desire* to be safe. If, however, safety is actually determined more by the motivation to be safe than by the physical opportunity that is offered or by the level of skill, the introduction of accident countermeasures of the engineering, education and enforcement varieties would not necessarily reduce the accident rate per head of population. What may occur instead is *behavioural adaptation*, that is a change in behaviour that offsets the potential safety benefits, as has been discussed in a recent OECD report (1990). The greater opportunity for safety and the increased level of skill may not be utilized for greater safety but instead for more advanced performance: 'Behavioural adaptations of road users, which may occur following the introduction of safety measures in the transport system, are of particular concern to road authorities, regulatory bodies and motor vehicle manufacturers particularly in cases where such adaptations may decrease the expected safety benefit' (OECD, 1990).

The OECD report cites numerous examples of this phenomenon. Taxicabs in Germany equipped with anti-lock brake systems were not involved in fewer accidents than taxis without these brakes. Increases in the lane width of two-lane highways in New South Wales in Australia have been found to be associated with higher driving speeds; for every 30 cm of additional lane width, speed increased by 3.2 km/h for passenger cars, while truck speed increased by about 2 km/h for every 30 cm increase in lane width. An American study dealing with the effects of lane-width reduction found that drivers familiar with the road reduced their speed by 4.6 km/h and those unfamiliar by 6.7 km/h. In Ontario, it was found that speeds decreased by about 1.7 km/h for each 30 cm of reduction in lane width. In Texas, roads with paved shoulders as compared to unpaved shoulders were found to be associated with speeds at least 10% higher. Drivers have generally been found to move at a higher speed when driving at night on roads with clearly painted edge markings.

Since the publication of the OECD report, Finnish studies have reported on the effect of installing reflector posts along highways with an 80 km/h speed limit. Randomly selected road sections, which totalled 548 km, were equipped with these posts and compared with 586 km that were not. The fact that the installation of reflector posts increased speed in darkness will no longer come as a surprise, nor that there was not even the slightest indication that it reduced the accident rate per kilometre driven on these roads; if anything, the opposite happened (Kallberg, 1992). Carefully matched samples of drivers with or without studded winter tyres were compared on their moving speeds, prudence in negotiating curves and following distance. The second group drove significantly more slowly on slippery roads and on curves, and maintained following distances that were longer by as much as 11 m (Mäkinen *et al.*, 1994).

More recently still, an American study reported that 'air-bag equipped cars tend to be driven more aggressively and that aggressiveness appears to offset the effect of the air bag for the driver and increase the risk of death of others' (Peterson *et al.*, 1995).

That the strength of the desire to be safe is the dominant factor in accident avoidance in the face of everyday physical injury risks on the road and on the job is one of the main contentions of the risk homeostasis theory, as was noted above. The art of safety management, therefore, is to effectively reduce the target level of risk of people at work and on the road. It was also noted above that this theory focuses on the accident rate per time unit of exposure to risk, e.g. per hour of road use or per head of population per year. This definition of the accident rate should be clearly distinguished from the accident rate per kilometre driven (or the accident rate per unit of production in occupational settings). An improvement in one does not imply an improvement in the other. For instance, the traffic death rate per unit distance driven in the USA dropped by a factor of approximately seven between 1927 and 1987, while the traffic death rate per 100 000 inhabitants was approximately the same in these two years and did not show a downward trend over a period spanning almost two-thirds of a century (see Wilde, 1994). In an 18-year period of economic growth. Ontario experienced a reduction of the fatality rate per kilometre driven of about 50%, while the traffic death rate per capita actually increased (see Wilde, 1994).

Reduction of the rate of accidents per unit distance of mobility or production is, of course, a meaningful objective and according to RHT it can be achieved by those accident countermeasures that allow operators to achieve greater driving speed or more units of production while maintaining the same accident risk per time unit of driving or work (see Wilde, 1994). In this chapter, however, we have opted for a reduction in the accident rate per time unit of exposure (per person-hour at work or on the road, or per capita per year) as the primary goal for accident prevention. This objective is analogous to a health-promotion programme seeking to reduce the smoking-related death rate per head of population, not per cigarette smoked (or an intervention that aims at reducing the AIDS mortality rate per capita, rather than per sex act).

8.3.2 Factors determining the target level of risk

The target level of accident risk is determined by four categories of motivating (i.e. subjective utility) factors:

1. The expected *advantages* of comparatively *risky* behaviour alternatives – for example gaining time by speeding, making a risky manœuvre to fight boredom.

2. The expected *costs* of comparatively *risky* behaviour alternatives – for example automobile repair expenses, insurance surcharges for being at fault in an accident.

3. The expected *benefits* of comparatively *safe* behaviour alternatives – for example an insurance discount for accident-free driving.

4. The expected *costs* of comparatively *safe* behaviour alternatives – for example using an uncomfortable seatbelt, being called a wimp by one's peers.

The higher the values in categories 1 and 4, the higher the target level of risk, while the target level of risk will be lower as the values in categories 2 and 3 rise. Some of the motivating factors in all four categories are economic in nature, others are cultural, social or psychological. They are usually so thoroughly internalized that most people are not consciously aware of them most of the time. Thus, the target level of risk should not be viewed as something that people arrive at as the result of explicit calculations of probabilities of various possible outcomes and the respective positive or negative values of these possible outcomes. A person who lowers the thermostat before going to sleep or when leaving home for the weekend usually does not choose a setting on the basis of precise calculations of expected cost and benefits, but rather by intuition. The same holds the next morning or after returning from the weekend trip when the person resets the target temperature on the home thermostat. The expression 'the target level or risk' should therefore not be understood as implying that people strive for a certain level of risk for its own sake. Target risk does not mean risk for the sake of risk, just as the target temperature set on a thermostat is not necessarily the choice of preference if energy costs were less important. Fever may well be useful in the body's fight against disease, but that does not mean that a fever is what you really want. Similarly, workers may reluctantly accept having to work under dangerous conditions for fear of losing their jobs.

It is obvious that economic motives play an important role among the factors that influence anybody's target level of risk. To move oneself or goods from A to B is a way of making money; to drive fast is to gain time but also means higher fuel costs and more vehicle wear and tear. In fact, the ups and downs in the economy have proven to be a very powerful predictor of the year-to-year fluctuations in the per capita traffic accident rate (Partyka, 1984; Adams, 1985; Zlatoper, 1989; Wilde, 1991, 1994; Farmer, 1996; Wilde and Simonet, 1996). Accident risk, however, may also be accepted for the purpose of satisfying other desires, such as seeking variety, curiosity, adventure and fighting boredom.

8.3.3 Incentives versus disincentives

The notion that the incidence of unwanted behaviour may be repressed by authorities acting on people's motivation has, of course, a long history, as is clear from the universal presence of punitive law. Although the threat and meting out of punishment for violations of the traffic code (in the absence of an accident) is also one of society's traditional attempts at motivating people towards safety, the evidence for its effectiveness still remains to be delivered (OECD – Road Research, 1974; Carr *et al.*, 1980; Bonnie, 1985). Even in cases in which selective enforcement does reduce the rate of a particular type of accident (e.g. drinking-and-driving accidents), this does not necessarily lead to a reduction in the overall accident rate because the rate of accidents with other immediate causes may increase (Wilde, 1988a, 1990). Sobriety is no guarantee for safety, neither is seatbelt use, obeying the speed limit or wearing safety boots. To believe that the overall accident rate will go down commensurably with the reduction in one particular immediate cause of accidents is to suffer from the delta illusion (Wilde, 1994). By obstructing channels (immediate accident causes or symptoms) one cannot reduce the amount of water that flows through a river delta to the ocean. To stop or reduce the flow, the solution has to be found upstream. In other words, one cannot reduce the accident rate by piecemeal measures that fail to affect the superordinate cause (the root cause, i.e. accident risk acceptance). Symptom substitution or 'accident metamorphosis' is the likely result of such interventions, not accident reduction.

On the other hand, there is evidence for the general deterrent effect of punishing road users for being involved in an accident in which they are at fault (e.g. Barmack and Payne, 1961). The approach that attempts to deter people from *specific unsafe acts* (in the absence of an accident) suffers from several other problems, some of which have been identified in the context of organizational psychology (Arnold, 1989):

(a) The 'self-fulfilling prophecy' of attribution. Labelling people with undesirable characteristics may stimulate individuals to behave as if they had these characteristics. Thus, the very imposition of a speed limit may provoke some people to drive faster than they otherwise would.

(b) The emphasis is on process controls (i.e. on specific behaviours, such as using a piece of safety equipment or obeying the speed limit, instead of on the outcome: safety). Process controls are cumbersome to design and implement. Moreover, process controls can never be totally exhaustive, that is cover all undesirable specific behaviours of all people all the time.

(c) Punishment brings negative side-effects. Punishment creates a dysfunctional organizational climate: resentment, uncooperativeness, antagonism, sabotage. As a result, the very behaviour that was to be prevented may in fact be stimulated.

In contrast to the questionable usefulness of the deterrence approach that focuses on violations (in the absence of an accident) in serving their intended purpose and the attendant undesirable side-effects, the evidence clearly indicates that incentive programmes for remaining free of accidents have the effect for which they are intended as well as the positive side-effect of creating a favourable social climate (Steers and Porter, 1991). The term 'incentive' refers to a *pre-announced* gratification or bonus extended to workers or drivers on the specific condition that they do not have an accident caused by their own fault within a specified time period (or 'incubation period'). Typical design features of incentive programmes are:

- *Target groups*: industrial employees in manufacturing, construction, food processing, holiday resorts, mining, etc.; truck drivers, bus drivers, passenger car drivers.
- *Single vs. multiple target groups*: workers only, workers as well as lower/middle management.
- *Nature of bonuses*: cash, savings bonds, public praise, certificates of merit, merchandise, gift certificates, savings stamps, lottery tickets, extra holidays, free driver's licence renewal, insurance rebates, insurance discounts.
- *Bonuses for individual vs. team performance*: for individual performance only, for team performance only, or a combination.
- *Condition for eligibility*: being accident-free, displaying specified safe behaviours, both.

- *Incubation periods*: one month, six months, one year.
- *Penalty for under-reporting of accidents*: present, absent.
- *Implementation*: incentive programme only or combined with other accident countermeasure (usually safety education).
- *Programme evaluation*: sometimes of high standard, sometimes deficient, sometimes absent.

The effectiveness of incentive programmes in enhancing safety has been clearly established (Komaki *et al.*, 1978; Wilde, 1985; Fox *et al.*, 1987; McAfee and Winn, 1989; Peters, 1991). In a comparatively recent meta-analysis of 53 published (in English only) evaluations of different types of occupational accident prevention, Guastello (1993) found that incentives were on average more effective towards safety than technological interventions, personnel selection and other types of intervention, including poster campaigns, near-accident reporting, quality circles and exercise and stress management.

8.4 COST-EFFECTIVENESS OF INCENTIVE PROGRAMMES

Reductions in the accident rate per employee-year down to fractions between 50% and 20% of base rate are not uncommon in manufacturing, construction and other industries. The transportation division of a German food processing plant saw a reduction in direct accident costs by more than two-thirds in the first year of implementing an incentive programme and the reduction remained at that level for over three decades (Gros, 1989). In other cases, the results are better still, as was the case in two American mining companies where the burden of lost days dropped by 89% and 98%, respectively (Fox *et al.*, 1987). Sometimes the results are more modest. A cable plant in the USA reduced the accident costs per employee by 35%, a manufacturer of tobacco products by 31% (Stratton, 1988), a grain processing and transportation company by 30%, a Pacific resort complex by 39% and a manufacturer of food products by a mere 10% (Bruening, 1989).

In some cases, major reductions in accident costs are achieved in return for low implementation costs of the incentive programme. On other occasions, the implementation costs may be comparatively high. Similarly, a minor reduction in accidents may sometimes be accomplished with an incentive programme that costs very little relative to the savings it produces. It is, therefore, of interest not only to consider the degree of accident reduction that may be achieved with incentive programmes but also the expense of the programme relative to the savings.

The ratios between benefits (savings on accidents prevented) and programme costs are usually greater than 2 to 1, meaning that industrial companies can make money on such accident prevention efforts (largely due to the reduction in fees to workers' compensation boards and other insurance that follows an improvement in a company's safety record). This favourable result raises a novel issue of equity: how are the profits to be divided between the owners of the company (whose managers put the incentive programme into place) and its employees (who deliver the increase in safety)?

The favourable effects last over time. Incentive plans in two American mines were studied over periods of 11 and 12 years. In one mine the number of days lost due to accidents was reduced by about 89% of baseline and in the other by as much as 98%. Benefit/cost ratios varied from year to year between 18 and 28 at one mine and between 13 and 21 at the other. There was no sign that the effectiveness of the incentive plans diminished over time at either mine (Fox *et al.*, 1987). A high benefit/cost ratio, about 23 to 1, has also been observed for incentives for safety in the resort hotel business (Bruening, 1989). An incentive programme implemented in a manufacturing plant of conveyor systems in Tennessee showed a benefit/cost ratio of 20 to 1 while bringing about a 77% reduction in accidents (Hatcher, 1991). Another programme was implemented in a construction company at a cost of about US$30000 a year and produced savings in workers' compensation insurance premiums of about US$400000 a year, which amounts to a benefit/cost ratio of approximately 13 to 1 (Synnett, 1992).

In passing, it may be noted that very favourable benefit/cost ratios have also been observed where incentives are used to combat absenteeism. Kelly (1982) reports a case of a ratio of about 26 to 1 over a 12-month period in a US electronics manufacturing plant. Another study, of transit bus drivers, showed a benefit/cost ratio of 5 to 1 and netted the transit company about US$175000 a year (Kelly, 1988).

Incentive programmes, with very few exceptions, are approved by the people to whom they are addressed and in this respect they compare favourably with the much less popular action of the law and of the police (Wilde, 1977). To put it popularly: a small carrot is not only much better liked than a big stick, it is also much more effective. Only one negative side-effect has been noticed so far, namely the tendency of people to under-report accidents when incentive programmes are in effect. Fortunately, however, such under-reporting has been found to occur with respect to minor accidents only (McAfee and Winn, 1989).

8.5 FACTORS ENHANCING THE EFFECTIVENESS OF INCENTIVE PROGRAMMES

Past experience with incentive programmes shows that some programmes have had much greater effects than others. For instance, a German incentive plan, which promised professional truck and van drivers a bonus of 350DM (at the time about US$160) for each half-year of driving without being at fault in an accident, produced a reduction in direct accident cost to less than one-third in the first year of application and remained at that level for over three decades (Tschnernitschek, 1978; Gros, 1989). In the California 'good driver' experiment, in which drivers in the general population were offered free extension of their driver's licence by one year in return for each year of accident-free driving, the accident rate dropped by 22% in the first year of the programme (incidentally, the accident reduction was the greatest for drivers under 25 years; Harano and Hubert, 1974). It is, therefore, important to identify the distinctive features of the more successful incentive schemes.

An effort has been made to cull the ingredients of the most effective incentive plans from the separate published reports. The bibliographic references to these may be found in Wilde (1985), McAfee and Winn (1989) and Peters (1991). Identification of the programme components that enhance the effectiveness of incentives has by necessity been an effort largely based on inference because, to date, there are no well-controlled experiments in which one particular incentive characteristic is varied and all other factors are kept constant. For obvious reasons, such experiments are not likely to be forthcoming; industry is not in the business of running such experiments. However, the following conditions would seem to favour the items that appear in the 'checklist' incentive effectiveness:

- strong managerial vigour and commitment;
- rewards are focused on not having an accident;
- high attractiveness of the reward;
- progressive accumulation of safety credits;
- simple rules;
- bonuses are perceived as equitable;
- bonuses are perceived as accessible;
- short incubation periods;
- enhanced peer pressure towards safe conduct;
- programme planned in consultation with the target population;
- under-reporting of lesser accidents is discouraged;
- incentives are extended to multiple levels of the organization;
- supplementing incentives with safety training is considered;
- maximization of net saving versus maximizing benefit/cost ratios; and
- provision for a research and development component.

These conditions will now be presented in more detail.

8.5.1 Strong managerial vigour and commitment

The introduction and long-term maintenance of incentive programmes should be conducted with managerial vigour, commitment and coherence. Workers or drivers should not only be informed of the programme in existence but they should also frequently be reminded of it in attention-catching ways. In order to motivate and inform the relevant audience those in charge of incentive programmes should provide clear and frequent knowledge of results to the target audience, i.e. *feedback* (Komaki *et al.*, 1978).

8.5.2 Rewards are focused on not having an accident

Incentive programmes should reward the outcome variable, i.e. the 'bottom line' of being free from accidents, not some process variable like wearing the seatbelt or safety glasses *per se*, driving when sober, obeying the speed limit or the rules regarding hours of service, or the wearing of a hard hat. This is because rewarding specific behaviours does not necessarily strengthen the motivation towards safety, and a potential safety benefit due to an increased frequency of one specific form of 'safe' behaviour may simply be offset by road users less frequently displaying other forms of 'safe' acting. 'The risk here is that while the rewarded behavior may improve, other related safe behaviors may deteriorate' (McAfee and Winn, 1989). This is why the evidence from incentive programmes for the purpose of increasing the frequency of particular safety behaviours – and there have been many such programmes – has not been incorporated in the present report. It can be found elsewhere (e.g. Geller, 1996).

One misgiving that has been expressed regarding the use of the 'bottom-line' criterion for eligibility to the bonus is the fact that chance plays a considerable role in the occurrence or non-occurrence of an accident (Vogel, 1991). Although risky behaviour and accident likelihood are clearly related in large samples of people. some individuals may chronically display unsafe behaviours yet get away with it because they are lucky. Others who are usually very prudent yet negligent on very rare occasions only may be unlucky enough to get involved in an accident on one of those occasions. Luck or chance, good or bad, may play an even bigger part if receiving the reward is made dependent on a lottery draw (section 8.5.7). In order to deal with this issue, a German manufacturer provided an example of a two-tiered criterion. Each month every employee who was not involved in an accident was eligible for a 100 DM cash draw but the employee whose name was drawn would receive the prize only if prior to the announcement of the winner she or he fulfilled another requirement as well. The other requirement demanded that the workstation was to be found clean and orderly and the employee seen to make good use of available safety equipment during an unannounced visit by the safety engineer and other company personnel (Bacher, 1989).

8.5.3 High attractiveness of the reward

Incentive programmes can be expected to be successful to the extent that they widen the utility difference between the perceived benefit of not having an accident and the perceived disadvantage of having an accident. Rewards for accident-free operation in industry have taken many different forms, ranging from cash to public commendation. They include trading stamps, lottery tickets, gift certificates, savings bonds, shares in company stock, extra holidays (Karasina, 1977) and other privileges. While the flexible uses of money prevents satiation from occurring, merchandise, especially customized merchandise, may constitute a lasting reminder of the value of safety. Merchandise items also have a 'value-added' component in the sense that they can be obtained at a lower price than the recipients would be likely to have to pay if they bought the items at retail. In the US, a substantial industry has sprung up to provide the merchandise for safety prizes. Gift certificates hold a middle ground between cash and merchandise; they can be put to flexible use and yet be personalized and imprinted with a commemorative message. The use of savings bonds has the advantage for the employer that they can be purchased at a price much below their value at maturity. Drivers have been rewarded with cash (Tschnernitschek, 1978; Gros, 1989), automobile insurance rebates (Vaaje, 1991) and free licence renewal (Harano and Hubert, 1974).

Awards do not have to be very large to be effective. In fact, a case can be made for relatively small awards being preferable in some cases. Small awards make it possible to hand out awards more frequently, they are probably less conducive to under-reporting of accidents and they may foster the internalization of pro-safety attitudes through the process of cognitive dissonance reduction (Geller, 1990). When a small reward changes a person's behaviour, that person may justify that change by reasoning that the change was for safety's sake rather than due to the insignificant inducement. No such internalization of pro-safety attitudes is necessary when the external inducement is large because in that case it may be perceived as fully justifying the behaviour change.

It should be noted, however, that the attitude-shaping effect of modest awards can only take place *after* the operators have changed their behaviour for whatever minor external inducement. The award should therefore be big enough to achieve some behaviour change to begin with.

In some cases a small material reward might imply a major social reward because of its 'symbolic function' (Markus, 1990). Safe behaviour may thus become the 'right thing to do'. This might help explain why a modest incentive such as free licence renewal for one year produced a major reduction in the accident rate of California drivers (Harano and Hubert, 1974). Moreover, analogous to earlier studies that found that increments in wages for dangerous work were related to increases in accident rate to the power of three, it may be suggested that small increments in wages for having no accidents should reduce the accident rate by a large amount (Starr, 1969).

8.5.4 Progressive accumulation of safety credits

The amount of the incentive should continue to grow progressively as the individual operator accumulates a larger number of uninterrupted accident-free periods, e.g. the bonus for 10 uninterrupted years of accident-free driving should be greater than 10 times the bonus for one year of accident-free driving (Wilde and Murdoch, 1982).

In Ontario, it is common practice for insurance companies to give fee discounts that are greater as the number of claim-free years increases. However, this is true only for up to five years of accident-free driving, although there is no reason for assuming that further discounts would have no accident-reducing effect beyond that period. What may be worse from the point of view of accident prevention is that a private motorist with five or more fault-free years who has an accident in which he or she is at fault is not likely to actually incur an increase in insurance fees. The reason the insurance companies have this 'forgiveness clause' would seem to be that the driver in question is seen as a relatively good risk whose business would be sadly missed if it went to the competition. This practice, however, fails to bring the accident rate down to a level as low as it perhaps could be otherwise (Wilde and Murdoch, 1982).

8.5.5 Simple rules

The operational rules of the programme should be kept simple so that they are easily understood by all persons to whom the programme applies. In the area of incentives for the purpose of increasing productivity it is known that too complex a set of rules may reduce programme effectiveness (Doherty *et al.*, 1989).

8.5.6 Bonuses are perceived as equitable

The incentive programme should be perceived as equitable by those to whom it is addressed. The bonus should be such that it is viewed as a just reward for not causing an accident in a given time period. Similarly, incentive systems should be designed such that those workers who are not eligible for the (top) award do not resent this and that those who are rewarded are seen by others as justly receiving the award (Markus, 1990). As chance plays a part in having or not having an accident, the actual receipt of the award may be made to depend on the additional requirement that the accident-free worker in question also maintains cleanliness and safety in his or her workstation (Bacher, 1989; section 8.5.2). In the event that disincentives are used as well, it is necessary that the target audience view the penalty imposed as justified.

8.5.7 Bonuses are perceived as accessible

Programmes should be designed such that the bonus is viewed as realistically attainable. This is of particular importance if the bonus is awarded in a lottery system. Lotteries make it possible to hand out greater awards, and this may enhance the attention-getting appeal of an incentive programme, but fewer among the people who have accumulated the safety credit will receive the bonus. This, in turn, may discourage some people from making an active attempt to accumulate the safety credit to begin with (Bartels, 1976).

8.5.8 Short incubation periods

The specified time period in which the individual has to remain accident-free in order to be eligible for the bonus should be kept relatively short. Delayed rewards and penalties tend to be discounted and are thus less effective in shaping behaviour than are more immediate consequences. Periods as short as one month have been used in industry. In the cited California experiment, those drivers whose licences were coming up for renewal within one year after being informed fo the incentive programme showed a greater reduction in accident rate than drivers whose licences were not to be renewed until two or three years later (Harano and Hubert, 1974).

8.5.9 Enhanced peer pressure towards safe conduct

Incentive programmes should be designed such that they strengthen peer pressure towards the objective of having no accidents. Thus, the plan should not only stimulate each individual worker or driver's concern for her or his own safety, but also motivate her or him to influence peers so that their accident likelihood is also reduced. In industrial settings this is achieved by extending a bonus for accident-free performance of the *work team* in addition to the bonus for individual freedom of accidents, and the team bonus has been found to increase the competitive motivation towards winning the team award. A dual bonus plan (individual *cum* team) for drivers in the same age bracket and living in the same city has been suggested and became known as the 'Saskatchewan plan' (Wilde, 1985). Team awards add a material incentive to act as one's brother's keeper and also have been found effective in isolation, that is in the absence of awards for individual performance (Vogel, 1991).

Social inducement towards safe conduct can also be enhanced by informing families of the safety award programme, the safety goals and potential rewards (Morisey, 1988) and by bonuses such as savings bonds or trading stamps that can be displayed in the homes of workers or drivers and remind them and their families that they were earned in recognition of safe operating performance (Fox *et al.*, 1987; Kirk, 1990).

8.5.10 Programme planned in consultation with the target population

The incentive scheme should be developed in cooperation and consultation with those people to whom it will be applied. People are more likely to actually strive for goals they themselves have helped define (Latham and Baldes, 1975; Komaki *et al.*, 1978). It should not be overlooked that an incentive can be an incentive only if the recipients view it as an incentive. Moreover, members of the target audience(s) are themselves likely to be the most informative as to which particular incentives are the most motivating towards increasing their safety performance under what conditions of perceived equity, accessibility, incubation and other features of any incentive programme under consideration. If the wishes and perceptions of the target audience are not considered in programme design, incentive plans may perform well below their potential (Gregersen *et al.*, 1996).

8.5.11 Under-reporting of lesser accidents is discouraged

Thought should be given to the question of how to counteract employees' tendency not to report the accidents they do have. That the institution of incentive progammes may stimulate this tendency seems to be the only currently identified negative side-effect of such programmes (while occasionally moral objections have been raised against rewarding people for obtaining a goal they should aspire to on their own, without being 'bribed into safety'; Hale and Glendon, 1987). Some incentive programmes have clauses providing for deduction of safety credits when accidents are not reported (e.g. Fox *et al.*, 1987). Fortunately, only those accidents that are minor remain unreported occasionally. However the greater the safety bonus, the more frequent this unreporting, including more serious hit-and-run accidents, may become.

8.5.12 Incentives are extended to multiple levels of the organization

Not only are shopfloor workers to be regarded for safe performance, but so are their supervisors and middle managers. This creates a more cohesive and pervasive safety orientation within a company (Zohar, 1980; Fox *et al.*, 1987; Bacher, 1989; Bruening, 1989; Synnett, 1992). Thus, several links in the line of command should be made eligible for an award, all the way up from worker to foreman, to supervisor and into middle management at least.

In the case of trucking safety, for instance, this requirement takes on special significance because there are reasons for believing that risk-taking inclinations of American and Canadian truck drivers are very much influenced by other officers in the companies in which they are employed, and by dispatchers in particular (Rothe,

1991). It would seem desirable to include them as potential bonus recipients in any incentive scheme aimed at improving driver safety.

8.5.13 Supplementing incentives with safety training is considered

Although educating towards safety is different from motivating towards safety, and a person's ability to be safe should be clearly distinguished from that person's willingness to be safe, some authors have expressed the feeling that it may be helpful to safety if workers are better informed of the specific behaviours through which accidents can be avoided (Tschnernitschek, 1978; Doherty et al., 1989; Peters, 1991).

8.5.14 Maximization of net savings versus maximizing benefit/cost ratios

In the planning of an incentive programme thought should be given to the question of what actually constitutes its primary goal: the greatest possible net savings due to accident reduction or a maximal benefit/cost ratio. Some programmes may reduce the accident frequency only slightly but achieve this at a very low cost. The benefit/cost ratio may thus be higher than is true for another progamme in which the ratio between benefits and costs is lower but which is capable of producing much greater net savings.

Consider the following example. Safety programme A can save US$700 000 at an implementation cost of US$200 000. Programme B can save US$900 000 at a cost of US$300 000. In terms of benefit/cost A's ratio is 3.5 while B's ratio equals 3.0. Thus, against the benefit/cost criterion, A is superior but if net savings are considered the picture is different. While programme A saves US$700 000 minus US$200 000, i.e. US$500 000, programme B saves US$900 000 minus US$300 000, i.e. US$600 000. In terms of net savings, the more expensive programme (B) is to be preferred.

Instead of adopting either net savings or the benefit/cost ratio as the criterion for choosing between programme options, one might favour a rather different objective: the greatest possible reduction in the accident loss that can be achieved on the sole condition that the costs of the programme do not to exceed the benefits. As an example of this, programme C saves US$1 000 000 at a cost of US$999 999. The benefit/cost ratios, as well as the net savings of C, are much lower than is true for either A or B but the savings are the greatest and exceed the costs.

8.6 EVALUATION ISSUES

Like any other accident countermeasure, an incentive plan should not be introduced without prior research into its short- and long-term feasibility, nor without prior research into its best possible form, nor without provision for scientifically adequate evaluation of its implementation costs and its observed effectiveness in reducing the accident rate (the final point on p. 93). The knowledge base of the safety research and application community is unlikely to grow without proper evaluation and ready access to publications.

The need for this may be illustrated by the following occurrence. Although it may seem improbable for a safety reward to actually have a negative effect on the subsequent safety record, there is one variation of a series of California reward/incentive programmes for the general driving public that produced worse driving records. Without programme evaluation, the surprising effect of this particular reward programme would never have come to light. In this programme, a benefit was given to drivers with no accidents on their records without their prior knowledge of that benefit. It took the form of an *unexpected reward* rather than an incentive, which highlights the importance of the distinction for safety promotion. The term 'incentive' refers to a *pre-announced* gratification or bonus extended to workers or drivers *on the specific condition* that they do not have an accident of their own fault within a specified time period. Unfortunately, the programme evaluation research did not go far enough to identify the reasons for the counterproductive effect of the reward condition. It is clear, however, that rewards and incentives should not be confused, nor their effects be assumed to be similar.

The simplest form of determining the effect of an incentive programme is a straightforward before/after design. Data are collected pertaining to the period before the introduction of the incentive plan and these are compared with data relevant to the situation following the introduction of the plan. There is, however, a possibility that the comparison is made invalid due to confounding factors that likewise changed from the one period to the other, e.g. changes in the economy, technology or legislation. This problem may be overcome by cancelling the incentive scheme some time after its introduction to see if this is followed by an increase in the accident rate. This alternative, however, may not be very attractive to a company because it would have to forego the programme's benefits.

Some researchers in the field of applied research, therefore, prefer a design that goes under the name of 'multiple-baseline study' (e.g. Fox *et al.*, 1987). This involves implementing an incentive scheme in different companies, or at different branches of the same company, *at different starting dates*. Showing evidence for the effectiveness of the scheme then takes the form of demonstrating that the drop in accidents indeed occurred in association with the date of intervention, while assuming that any confounding factors would be equivalent for the different companies (or different branches).

The following research designs are available. If we call the condition without the incentive plan A, while B refers to the situation in which the incentive plan is in effect, the multiple-baseline design may be referred to as the staggering over time of A–B–B in one group of employees in synchrony with A–A–B in another. A simple before/after comparison with external control data would take the form of A–B in the incentive group and A-A in the controls. Implementing the incentive programme and then withdrawing it after some time would take the form of A-B-A in one group and A-A-A in the other.

Increased safety is, by definition, the criterion of the effectiveness of a safety incentive programme. However, it may be assessed in various ways, e.g.:

- reduction in doctor's visits occasioned by accidents;
- reduction in days lost;
- reduction in material damage;
- reduction in insurance costs;
- reduction in time delays or down-time due to accidents; and
- reduction in recruiting and training costs of replacing operators.

Moreover, the available experience related to incentive plans shows that these may have side-effects, both desirable and undesirable. Thus, there is reason for including among the dependent variables:

- the rate of under-reporting of minor accidents;
- productivity; and
- the rate of personnel turnover.

A further interesting endeavour for future research may be to determine what individuals actually do or refrain from doing in order to attain the anticipated bonus. How is their driver behaviour altered? The literature on incentives in industry makes occasional mention of workers keeping their workstations cleaner and more orderly after an incentive programme has been put in place. Preliminary data from an experimental study of incentives on driver behaviour of private motorists in The Netherlands indicate that they reduce speed and increase following distance (Heino *et al.*, 1996).

8.7 TARGET AUDIENCES: IN-PLANT WORKERS *VERSUS* DRIVERS

Although incentives for accident-free performance have been shown to be effective in making industrial workers as well as drivers of private vehicles and company vans and trucks behave more safely, there are differences between the three settings of operation that have implications for programme design and application. For one thing, workers usually operate in relatively small teams of people who know each other relatively well. Drivers of motor vehicles operate in a situation of near-anonymity. In several ways, truckers and public-transit drivers may occupy a position that is intermediate between the one of shopfloor workers and drivers of private cars. Workers in industry occupy positions within a clear line of command, while there is no formal hierarchy between passenger-car drivers beyond the context of the highway code. It is thus more difficult, though not impossible (Wilde, 1988b), to design an incentive system that enhances peer pressure towards safe conduct.

In industry it is relatively easy to keep the incubation period of the award quite short (e.g. as short as a month, like the interval between pay cheques); for drivers in the general population this might be unmanageable for administrative reasons. More importantly perhaps, it is obviously advantageous for companies to institute incentive systems because the savings (including discounts in insurance fees) are usually very much greater than the implementation costs. The most attractive incentives to drivers in the general population are likely to be those that could in principle be offered by automobile insurance companies. This could take the form of tangible discounts or rebates for claim-free driving. On the other hand, the very fact that insurance can be bought against certain hazards diminishes the threat of the consequences of these hazards and, therefore, may be expected to increase people's willingness to expose themselves to these hazards, with an increase in accident losses as a consequence. It is not surprising that increases in workers' compensation payments for injuries may increase the rate of workplace accidents (Worrall, 1983; Fortin and Lanoie, 1992; Butler *et al.*, 1996; Loeser *et al.*, 1996). Phrased in simple terms: to offer people protection against the consequences of risky behaviour

encourages risky behaviour; to offer people still better protection against the consequences of risky behaviour encourages riskier behaviour still. The very existence
of motor-vehicle insurance may have the effect of increasing the size of the problem
it sells protection against. Not surprisingly, automobile insurance was at one time
forbidden by law in some parts of the world (Ewald, in press). For the sake of traffic
safety it is, therefore, desirable that insurance practices are structured in a manner
such that the risk-taking tendencies of the customers are effectively being discouraged (Wilde, 1994, in press).

8.8 CONCLUSION

The argument laid out in this chapter may be conveniently recapitulated as
follows:

- Of all accident countermeasures that are currently available, those that affect
  people's motivation towards safety seem to be the most promising.
- Of all countermeasures that affect people's motivation towards safety, those that
  reward people for accident-free performance seem to be the most promising. To
  maintain that safety is its own reward is to ignore the fact that people knowingly
  engage in risky behaviours in anticipation of the benefits expected from those
  behaviours.
- Of all possible incentive schemes that reward people for accident-free performance, some promise to be more effective than others because they contain the
  elements that appear to enhance motivation towards safety.

The literature review by McAfee and Winn contains the following statement:
'The major finding was that every study, without exception, found that incentives or
feedback enhanced safety and/or reduced accidents in the workplace, at least in the
short term. Few literature reviews find such consistent results' (McAfee and Winn,
1989).

There can be no question that incentive programmes are the most powerful
safety tools for safety on the job and on the road. The question therefore is, why
are they not much more frequently implemented than is already the case? One
factor may be ignorance of the existing know-how. Another factor may be the
fear that accident under-reporting will be the result. Although there is some
justification for this, in industry it may be possible to conceal a minor injury but
one cannot hide a corpse. There may be resistance because of the attitudes of
unions, of management or, more generally, company climate (Petty et al., 1992).
Specific incentives that are taken seriously by operators in one cultural context
may be viewed as mere 'hoopla' in another (Lonero and Wilde, 1992) or even
as childish (Hagenzieker, 1992). On the other hand, their implementation would
seem to lead to satisfaction in the large majority of companies that have tried
them; 84% of a random sample of manufacturing companies in Wisconsin that
used incentive programmes reported a reduction in accidents and 70% deemed
the programmes cost-effective (Lundblad, 1985). Note that the expertise in
incentive scheme design has become considerably more sophisticated since that
time and that more up-to-date programmes would be expected to lead to even
better results.

Besides having possible negative side-effects, incentives also have *positive* side-effects. For one thing, incentives are a money-making proposition in industry: savings usually exceed the reduction in accidents by a large degree, as has been discussed above. For another, they may help develop to better company morale (Fox *et al.*, 1987). As is true for successful productivity gainsharing programmes (Doherty *et al.*, 1989), safety incentive programmes can help improve the general organizational climate and, therefore, make a positive contribution to productivity over and above the gain due to accident reduction. Safety incentives give workers a common cause with each other as well as with management. Reinforcing safe acts 'removes the unwanted side-effects with discipline and the use of penalties; it increases the employees' job satisfaction; it enhances the relationship between the supervisor and employees . . . ' (McAfee and Winn, 1989). Moreover, it has been observed that incentive programmes can provide people with benefits over and above the mere material rewards: '. . . I have seen employees develop positive attitudes, a sense of worth, stronger loyalties to their job and company and greater motivation toward safety – all as a result of a safety incentive program' (Sheldon, 1986).

References

ADAMS, J.G.U. (1985) *Risk and freedom: the record of road safety legislation.* London: Transport Publishing Projects.

ALEXANDERSSON, S. (1972) *Some data about traffic and traffic accidents.* Stockholm: The Swedish Road Safety Office.

ARNOLD, H.J. (1989) Sanctions and rewards: organizational perspective. In *Sanctions and rewards in the legal system: a multidisciplinary approach*, ed. M.L. Friedland, Toronto: University of Toronto Press.

BACHER, K. (1989) Erfahrungen mit dem Sicherheitswettbewerb 'Sicher arbeiten und 100 Mark gewinnen' bei der Hoogovens Aluminiums Hüttenwerk GmbH (Work without accidents and earn 100 DM in the Hoogovens aluminum plant). In *Psychologie der Arbeitssicherheit, 4, Workshop 1988*, ed. B. Ludborzs, pp. 345–346, Heidelberg: Roland Ansager Verlag (in German).

BARMACK, J.E. and PAYNE, P.E. (1961) The Lackland accident counter measure experiment. *Highway Research Board Proceedings*, **40**, 513–522.

BARTELS, K. (1976) *Über die Wirksamkeit von Arbeitssicherheitsprämien (Studies of the effectiveness of occupational safety incentive programmes).* Dortmund: Bundesanstalt für Arbeitsschutz und Unfallforschung (in German).

BONNIE, R.J. (1985) The efficacy of law as a paternalistic instrument. *Nebraska Symposium on Motivation*, **29**, 131–211.

BRUENING, J.C. (1989) Incentives strengthen safety awareness. *Occupational Hazards*, Nov., 49–52.

BUTLER, R.J., DURBIN, D.L. and HELVACIAN, N.M. (1996) Increasing claims for soft tissue injuries in workers' compensation: cost shifting and moral hazard. *Journal of Risk and Uncertainty*, **13**, 73–87.

CARR, A.F., SCHNELLE, J.F. and KIRCHNER, R.E. (1980) Police crackdowns and slowdowns: a naturalistic observation of changes in police traffic enforcement. *Behavioral Assessment*, **2**, 33–41.

COLLEDGE, A. (1995) To improve cost, improve morale. *Occupational Health and Safety*, **64**, 23–24.

DOHERTY, E.M., NORD, W.R. and McADAMS, J.L. (1989) Gainsharing and organization development: a productive synergy. *Journal of Applied Behavioral Science*, **25**, 209–229.

EWALD, F. (in press) Assurance et prévention: conflit ou convergences des logiques? *Proceedings, International Conference on Automobile Insurance and Road Accident Prevention, organized by the OECD Road Research Programme and the Dutch Ministry of Transport and Public Works*, Amsterdam, April, 6–8, 1992.

FARMER, C.M. (1996) *Trends in Motor Vehicle Fatalities*, Technical Report, Arlington, Virginia: Insurance Institute for Highway Safety.

FORTIN, B. and LANOIE, P. (1992) Substitution between unemployment insurance and workers' compensation: an analysis applied to the risk of workplace accidents. *Journal of Public Economics*, **49**, 287–312.

FOX, D.K., HOPKINS, B.L. and ANGER, W.K. (1987) The long-term effects of a token economy on safety performance in open pit mining. *Journal of Applied Behavior Analysis*, **20**, 215–224.

GELLER, E.S. (1990) Shaping workers' attitudes toward safety. *Occupational Hazards*, **52**, 49–51.

(1996) *The psychology of safety: how to improve behaviors and attitudes on the job*, Radnor, PA: Chilton.

GERSON, V. (1994) Incentive programs prove rewarding for employees who earn recognition. *Occupational Health and Safety*, **63**, 40.

GREGERSEN, N.P., BREHMER, B. and MORÉN, B. (1996) Road safety improvements in large companies: an experimental comparison of different measures. *Accident Analysis and Prevention*, **28**, 297–306.

GROS, J. (1989) Das Kraft-Fahr-Sicherheitsprogramm (The road safety programme of the Kraft company). *Personalführung*, **3**, 246–249 (in German).

GUASTELLO, S.J. (1993) Do we really know how well our occupational accident prevention programs work? *Safety Science*, **16**, 445–463.

HAGENZIEKER, M. (1992) Beloningen voor verkeersveilig gedrag (Rewards for safety in traffic). *De Psycholoog*, Sept., 349–353 (in Dutch).

HALE, A.R. and GLENDON, A.I. (1987) *Individual behaviour in the control of danger*, Amsterdam: Elsevier.

HARANO, R.M. and HUBERT, D.E. (1974) *An evaluation of California's 'good driver' incentive program*. Report No. 6, California Division of Highways, Sacramento.

HATCHER, E. (1991) Positive safety. *Training*, July 39–41.

HEINO, A., VAN DER MOLEN, H.H. and WILDE, G.J.S. (1996) Effect van beloningen voor veilig rijden op rijgedrag and risicoperceptie (The effect of safety incentives on driver behaviour and risk perception). In *Gedragsbeïnvloeding in verkeers- en vervoersbeleid*, eds F.J.J.M. Steyvers and P.G.M. Miltenburg, pp. 71–74, Groningen: Traffic Research Centre, University of Groningen.

HISLOP, R.D. (1993) Developing a safety incentive program. *Professional Safety*, **38**, 20–25.

KALLBERG, V.P. (1992) *The effects of reflector posts on driving behaviour and accidents on two-lane rural roads in Finland*, Helsinki: The Finnish National Road Administration Technical Development Center, Report 59/1992.

KARASINA, N.I. (1977) *Psychological and material incentives for the improvement of workplace conditions*, Moscow: Scientific Research Institute for Occupational Safety (in Russian).

KELLY, L. (1982) *Absenteeism: policies and programs for the 80s*, Kingston, Ontario: IR Research Services.

(1988) *Attendance incentive plans*, Kingston, Ontario: IR Research Services.

KIRK, P.F. (1990) Safety incentive plan. *Professional Safety*, **35**, 38–40.

KOMAKI, J., BARWICK, K.D. and SCOTT, L.R. (1978) A behavioral approach to occupational safety: pinpointing and reinforcing safe performance in a food manufacturing plant. *Journal of Applied Psychology*, **63**, 434–445.

LATHAM, G.P. and BALDES, J.J. (1975) The practical significance of Locke's theory of goal setting. *Journal of Applied Psychology*, **60**, 122–124.

LEGLER, J.A. (1992) Safety incentives produce measurable results. *Waste Age*, **23**, 67–68.

LOESER, J.D., HENDERLITE, S.E. and CONRAD, D.A. (1996) Incentive effects of workers' compensation benefits: a literature synthesis. *Medical Care Research and Review*, **52**, 34–59.

LONERO, L.P. and WILDE, G.J.S. (1992) Get your incentive program off the ground. *Occupational Health and Safety Canada*, **8**, 62–67.

LUNDBLAD, E.C. (1985) Incentive programs reduce accidents, save money. *National Safety News*, Jan. 35–37.

MÄKINEN, T., BEILINSON, L., RATHMAYER, R. and WUOLIJOKI, A. (1994) *The effect of studded tyres on journeys and driver risk taking*. Espoo: VTT Communities and Infrastructure, Transport Research Report 239.

MARKUS, T. (1990) How to set up a safety incentive program. *Supervision*, July, 14–16. (1991) Set up a safety incentive program. *Chemtech*, **21**, 350–351.

MAZZURCO, J.P. (1992) Creating a desire for safety. *Aberdeen's Concrete Construction*, **37**, 647.

MCAFEE, R.B. and WINN, A.R. (1989) The use of incentives/feedback to enhance work place safety: a critique of the literature. *Journal of Safety Research*, **20**, 7–19.

MCILWAINE, K. (1994) Compensation system law under review: New Zealand employers press for right to audit fund. *Business Insurance*, **28**, 17.

MORISEY, M. (1988) Award programs reduce costs, improve worker safety records. *Occupational Health and Safety*, Sept., 64–66.

OECD–Road RESEARCH (1974) *Research on traffic law enforcement*, Paris: OECD.

OECD–ROAD TRANSPORT RESEARCH (1990) *Behavioural adaptation to changes in the road transport system*, Paris: OECD.

PARTYKA, S. (1984) Simple models of fatality trends using employment and population data. *Accident Analysis and Prevention*, **16**, 211–222.

PETERS, R.H. (1991) Strategies for encouraging self-protective employee behavior. *Journal of Safety Research*, **22**, 53–70.

PETERSON, S., HOFFER, G. and MILLNER, E. (1995) Are drivers of air-bag equipped cars more aggressive? A test of the offsetting behavior hypothesis. *Journal of Law and Economics*, **37**, 251–264.

PETTY, M.M., SINGLETON, B. and CONNELL, D.W. (1992) An experimental evaluation of an organizational incentive plan in the electric utility industry. *Journal of Applied Psychology*, **77**, 427–436.

REDMANN, K.P. (1993) Chevron contractor incentives stress safety and environmental compliance. *Petroleum Engineer International*, **65**, 38–39.

ROTHE, J.P. (1991) *The trucker's world: risk, safety and mobility*. New Brunswick, NJ: Transaction Publishers.

SHELDON, J. (1986) How to set up a safety incentive program. *Occupational Health and Safety Canada*, **2**, 28–32.

STARR, C. (1969) Social benefits versus technological risk. What is our society willing to pay for safety? *Science*, **165**, 1232–1238.

STEERS, R.M. and PORTER, L.W. (1991) *Motivation and work behavior*, 5th edn, New York: McGraw-Hill.

STRATTON, J. (1988) Low-cost incentive raises safety consciousness of employees. *Occupational Health and Safety*, Mar. 12–15.

SYNNETT, R.J. (1992) Construction safety: a turnaround program. *Professional Safety*, **37**, 33–37.

TSCHNERNITSCHEK, E. (1978) Verkehrssicherheitsprogramm eines Vertriebsunternehmens (A company's traffic safety programme), *Berufsgenossenschaft*, February (in German).

VAAJE, T. (1991) Rewarding in insurance: return of part of premium after a claim-free period. *Proceedings, OECD/ECMT Symposium on enforcement and rewarding: strategies and effects*, Copenhagen, Denmark, Sept. 19–21, 1990.

VOGEL, C.B. (1991) How to recognize safety. *Safety and Health*, Jan., 54–57.

WEINSTOCK, M.P. (1994) Rewarding safety. *Occupational Hazards*, **56**, 73–76.

WILDE, G.J.S. (1977) A psychological study of drivers' concern for road safety and their opinions of various public policy measures against drinking and driving. *Proceedings, 7th International Conference on Alcohol, Drugs and Traffic Safety*, Melbourne, Australia, Jan. 23–28, pp. 410–424.

(1985) The use of incentives for the promotion of accident-free driving. *Journal of Studies on Alcohol*, supp. no. 10, 161–168.

(1988a) Risk homeostasis theory and traffic accidents: propositions, deductions and discussion of dissension in recent reactions. *Ergonomics*, **31**, 441–468.

(1988b) Incentives for safe driving and insurance management. In *Report of inquiry into motor vehicle accident compensation in Ontario*, vol. II, ed. C.A. Osborne, pp. 464–511, Toronto: Queen's Printer for Ontario.

(1990) Questioning the progress: the matter of yardsticks and the influence of the economic juncture. *Proceedings, 11th International Conference on Alcohol, Drugs and Traffic Safety*, Oct. 24–27, 1989, Chicago, IL.

(1991) Economics and accidents: a commentary. *Journal of Applied Behavior Analysis*, **24**, 81–84.

(1994) *Target risk: dealing with the danger of death, disease and damage in everyday decisions*, Toronto: PDE Publications.

(in press) Modification of driver behaviour through incentives for accident-free driving; implications for automobile insurance practices. *Proceedings, International Conference on Automobile Insurance and Road Accident Prevention, organized by the OECD Road Research Programme and the Dutch Ministry of Transport and Public Works*, Amsterdam, April 6–8, 1992.

WILDE, G.J.S. and MURDOCH, P.A. (1982) Incentive systems for accident-free and violation-free driving in the general population. *Ergonomics*, **25**, 879–890.

WILDE, G.J.S. and SIMONET, S.L. (1996) *Economic fluctuations and the traffic accident rate in Switzerland: a longitudinal perspective*, Technical Report, Berne: Swiss Council for Accident Prevention.

WORRALL, J.D. (1983) *Safety and the workforce: incentives and disincentives in workers' compensation*, New York State School of Industrial and Labor Relations, Cornell University: ILR Press.

ZLATOPER, T.J. (1989) Models explaining motor vehicle death rates in the United States. *Accident Analysis and Prevention*, **21**, 125–154.

ZOHAR, D. (1980) Promoting use of personal protective equipment by behavior modification techniques. *Journal of Safety Research*, **12**, 78–85.

CHAPTER NINE

Minimizing the risk of occupationally acquired HIV/AIDS: universal precautions and health-care workers

DAVID M. DEJOY, ROBYN R.M. GERSHON AND
LAWRENCE R. MURPHY

9.1 INTRODUCTION

Universal precautions (UP) are recommended work practices designed to protect health-care workers (HCWs) from exposure to blood-borne pathogens, such as HIV/AIDS and hepatitis B. Specific precautions include proper disposal of needles and other sharps, not recapping used needles and using disposable latex gloves and other protective garments and equipment. The US Centers for Disease Control and Prevention (CDC) issued formal guidelines related to UP in 1987 (Centers for Disease Control, 1987) and UP became mandatory in the USA in 1991 with the passage of the OSHA Blood-borne Pathogens Standard (Occupational Safety and Health Administration, 1991). These actions notwithstanding, a number of studies indicate that compliance with UP is inconsistent and often quite poor (e.g. Becker *et al.*, 1990; Kelen *et al.*, 1990, Hersey and Martin, 1994).

Thus far, strategies to prevent occupational exposures to blood-borne pathogens have focused on modifying worker behaviour and work practice controls. This chapter argues that the minimization of risk from blood-borne pathogens should proceed from a comprehensive or work-systems analysis of job/task, worker and environmental/organizational factors (Figure 9.1). The defining assumption of a systems or multi-level approach is that complex actions and events cannot be fully understood by examining them in isolation. The UP-related actions of individual workers should not be analyzed without detailed consideration of job demands and broader organizational and environmental influences. The available literature on UP, including recent findings from NIOSH-sponsored research, suggests the potential of such an approach and provides insight into the limited success of current

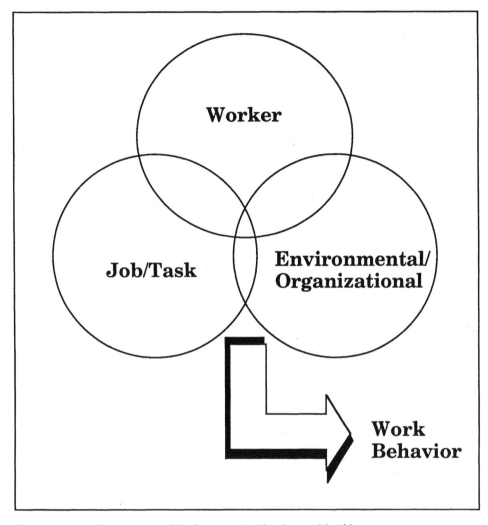

Figure 9.1   Work-systems model of occupational safety and health.

regulatory and related actions. Relevant findings are summarized in six areas: job demands and risk of exposure, safe work practices and job performance, knowledge and skills needed for hazard avoidance, attitudes and beliefs related to self-protective action, workplace design and organization, and organizational safety climate. The first two of these areas pertain to job/task factors, the second two to worker factors and the remaining two to organizational/environmental factors.

9.2 JOB/TASK DEMANDS AND RISK OF EXPOSURE

Job/task demands include the physical and psychosocial requirements of particular jobs, and a fundamental consideration is the extent to which job/task demands contribute to the risk of occupational exposure to blood-borne pathogens. Unfortunately, the UP literature contains relatively little information derived from direct analysis of specific patient care or other relevant tasks. Needlesticks are one exception to this conclusion and illustrate how careful analysis of specific jobs and tasks can provide useful insights. First, studies of needlestick injuries show that only between about 9 and 25% of needlesticks occur during recapping (Ruben *et al.*, 1983; Krasinski *et al.*, 1987; McCormick *et al.*, 1991; Yassi and McGill, 1991). Thus, prohibiting recapping, which is a centrepiece of UP, is relevant to only a fraction of needlestick injuries. Many needlesticks occur through handling or coming into contact with exposed needles and not during recapping *per se*.

Second, these studies show that strong countervailing forces exist to support continued recapping in some situations. For example, many older HCWs were originally trained to recap used needles as part of good safety practice. These workers are now being asked to change an established and well-rehearsed work practice. In addition, there may be valid reasons for recapping under certain circumstances. Some needled devices, such as certain types of intravenous tubing/needle assemblies, must be disassembled prior to disposal (Ribner *et al.*, 1987; Jagger *et al.*, 1988). By not recapping, the HCW is forced to handle or work in close proximity to exposed needles during disassembly.

Third, this research has provided a better understanding of injury mechanisms and high-risk tasks, which has accelerated the development of engineering controls. A variety of shielded and needleless systems have been developed and these innovations suggest that the handling of exposed needles can be reduced significantly or possibly eliminated in some situations (e.g. New York State Department of Health, 1992; Younger *et al.*, 1992).

Viewed together, these conclusions highlight the need for complementary behavioural and environmental actions. Priority should be given to making maximum use of available engineering technology to reduce direct exposure to exposed needles and to remove the incentive to recap. On the behavioural side, active prompts and reminders will be required to discourage recapping and to promote due care in situations where exposed needles cannot be eliminated. Management actions that clearly establish responsibilities for needle disposal and that emphasize the importance of this aspect of UP are also needed.

9.3 SAFE WORK PRACTICES AND JOB/TASK PERFORMANCE

Another important aspect of examining job/task factors concerns whether safe work practices and other control strategies interfere with job or task performance. Surveys of HCWs indicate that lack of time and interference with skilful task performance are the most frequently reported reasons for non-compliance (e.g. Kelen *et al.*, 1990; Hoffman-Terry *et al.*, 1992). Interference with the patient–practitioner relationship has also been mentioned in several studies (e.g. Linn *et al.*, 1990). In the NIOSH study, hospital-based HCWs who perceived a low level of conflict between job demands and self-protection were more than twice as likely to be compliant than those who perceived high levels of conflict (Gershon *et al.*, 1995). In subsequent analyses of these data (DeJoy *et al.*, 1995a), job hindrances were found to be the best predictor of non-compliance for both nurses and physicians. Physicians reported greater job-related hindrances and claimed to have less knowledge and information about UP than did either nurses or technicians; they also had the lowest overall compliance (Table 9.1).

Table 9.1 Group means and univariate Fs comparing nurses, physicians and technicians

Dimension	Nurses	Physicians	Technicians	F
Safety performance feedback	2.85a	2.69b	2.84a	7.88*
Personal protective equipment	3.40a	3.38a	3.42a	0.43
Job hindrances	1.71a	1.87b	1.65a	16.18*
UP knowledge and information	3.38a	3.02b	3.31a	49.89*
Compliance with UP	3.29a	3.13b	3.20b	14.34*

Adapted from DeJoy *et al.* (1995a). Safety performance feedback, personal protective equipment, job hindrances and UP knowledge and information are composite measures using five-point rating scales; compliance with UP is a composite measure using a four-point rating scale. For all measures, higher scores indicate higher levels of the dimension (greater job hindrances, better compliance, etc.). Means with the same letter are not significantly different.
*$p < 0.001$.

Two conclusions about UP practices and job performance seem clear. First, HCWs believe that UP requirements interfere with the optimal performance of their jobs. Second, in contrast to most other occupations, there is very little tolerance in health care for performance decrements associated with the use of personal protective equipment or any other hazard control measure. However, it is important to note that these conclusions come almost exclusively from HCW self-reports. There has been very little direct observation and analysis of patient-care activities. The patient perspective is also noticeably lacking.

9.4 KNOWLEDGE AND SKILLS PERTINENT TO HAZARD AVOIDANCE

Turning to the worker analysis portion of Figure 9.1, recent surveys suggest that most HCWs possess adequate knowledge of UP practices and occupational transmission of blood-borne pathogens. Hersey and Martin (1994), in their national survey, found that 89% of patient-care staff had attended at least one training session on infection control procedures. In the NIOSH study, approximately 75% of respondents had participated in some type of UP-related training and 95% were classified as having a high level of knowledge about UP (Gershon *et al.*, 1995). Still, knowledge by itself is seldom sufficient to produce or sustain safe behaviour. Although the large majority of HCWs in the Hersey and Martin survey had participated in training activities, only 43% 'always' wore gloves to draw blood, 55% recapped at least 'sometimes' and only 63% 'always' washed their hands after removing their gloves. In the NIOSH study, three-quarters of respondents had taken part in training but only about 24% were classified as 'fully compliant' (i.e. they 'always' or 'often' adhered to each of 11 UP-related practices). Compliance rates across the 11 behaviours ranged from about 97% for glove usage to 73% for not recapping and 56% for wearing disposable face masks.

Aggressive information dissemination and mandated HCW training have produced undeniable benefits in terms of improved awareness and knowledge levels. However, at this juncture, additional information-based training may yield diminishing returns in terms of producing further improvements in compliance. Skills- and strategies-oriented training may prove more useful, especially in view of the apparent importance of job hindrances in non-compliance. Before this can occur, additional information is needed concerning the perceived and actual skill levels of HCWs in practicing UP and in avoiding or managing high-risk situations.

9.5 ATTITUDES AND BELIEFS RELATED TO SELF-PROTECTIVE ACTION

Most theoretical models of self-protective behaviour assign considerable import-ance to the individual's threat-related beliefs and to the processing of cost-benefit information (DeJoy, 1996). Using a general health-belief model framework (e.g. Becker, 1974), Table 9.2 tries to organize what is known about the contribution of these factors to compliance with UP. With respect to threat-related beliefs, research suggests that most HCWs do not dismiss or underestimate their personal risk (susceptibility) of blood-borne infection (e.g. Linn *et al.*, 1990; Smyser *et al.*, 1990; Hoffman-Terry *et al.*, 1992). There is, however, some indication that HCWs may sometimes act on the basis of situation-specific as opposed to aggregate or overall risk. For example, Hoffman-Terry and colleagues reported that among medical and surgical residents the most frequent reason for not reporting exposures was that they did not view the specific exposure to be a significant health risk. Concerning perceived severity, there is little disagreement about the seriousness of the diseases that can be transmitted through exposure to contaminated blood and other body fluids.

Moving to outcome-related beliefs, HCWs seem to possess a reasonable degree of confidence in the effectiveness of UP as a preventive measure. Kelen and associ-ates (Kelen *et al.*, 1990) found that only 2.7% of emergency room personnel felt that UP do not work. Hoffman-Terry and associates (Hoffman-Terry *et al.*, 1992) found that 97% of medical and 69% of surgical residents strongly disagreed with the statement 'precautions are ineffective'. Gershon and colleagues (Gershon *et al.*, 1995) found that 95% of hospital workers agreed with the statement 'if UP are followed, my risk will be low'.

As discussed earlier, job-related barriers do appear to be an important factor in non-compliance. Patient care is a unique type of work activity. To begin with, the 'product' is human health itself and, by tradition, the patient's needs come first. In addition, many medical care situations afford precious little tolerance for anything that is thought to delay or encumber treatment in any manner. Another important

Table 9.2   Summary of health-care worker beliefs related to compliance with UP

Belief category	Research findings support compliance		
	Yes	No	Unclear
Threat-related			
Perceived susceptibility	X		
Perceived seriousness	X		
Outcome-related			
Perceived benefits	X		
Perceived barriers		X	
Efficacy expectations			
Perceived ability to comply			X

factor is that unlike almost all other work situations involving personal protective equipment, UP practice has an obvious interpersonal dimension. Instead of simply providing a barrier between the worker and some environmental hazard, UP place physical barriers between two interacting and essentially interdependent people.

Finally, relatively little is known about the role of self-efficacy in the compliance process. Self-efficacy is the extent to which HCWs believe that they can successfully perform UP behaviours and comply with UP guidelines. The potential link between self-efficacy and job hindrances requires further study. An important leverage point for improving compliance may rest with altering the self-efficacy expectancies of HCWs through skill- and strategies-based training. In fact, further efforts to increase overall levels of perceived susceptibility may actually be counter-productive without commensurate attention to self-efficacy enhancement and/or barrier reduction (Job, 1988).

9.6 WORKPLACE DESIGN AND ORGANIZATION

The organizational analysis involves a broad-based examination of the physical and social factors that transcend specific jobs. The need for both micro- and macro-task analyses of medical care environments has been pointed out by several authors (e.g. Casey, 1993; Bogner, 1994) but, to date, relatively little systematic research attention has been devoted to this issue. Most medical care settings involve groups of specialized and interdependent workers interacting with each other and with various types of equipment and devices. These settings are complex and dynamic in terms of both person–technology and person–person interactions. In such environments, safety performance can decline in a non-linear fashion as workload and situational demands increase.

Kelen and colleagues (Kelen et al., 1990) found that in the emergency department at a large medical centre compliance with UP was 44.7% for situations involving no bleeding, 57.7% with active bleeding but only 19.5% in the presence of profuse bleeding. Similarly, compliance was only 16.7% for major interventions, as compared to 56.4% for minor interventions and 44.1% for exams. Hammond and colleagues (Hammond et al., 1990) found a strict compliance rate of only 16% among surgical residents engaging in trauma room resuscitation. Even for highly invasive procedures, such as inserting chest tubes, compliance was less than 40%. These findings suggest that adherence may sometimes be poorest when the risk of exposure is greatest. The identification and analysis of special compliance requirements and high-risk situations should be an important feature of a comprehensive infection control programme. It is very important to recognize that the total risk faced by any HCW is influenced by situational factors and by the actions taken or not taken by other workers. It is both incorrect and unfair to assume that HCWs have total control over their own compliance behaviour.

9.7 ORGANIZATIONAL SAFETY CLIMATE

Safety climate refers to the perceptions that workers share about safety in their organization (Zohar, 1980). The importance of climate-like factors has been discussed with respect to health care in general (Cox and Leiter, 1992) and UP in particular (White and Berger, 1992). Safety climate is generally thought to provide a frame of reference for developing coherent expectations about behaviour–outcome contingencies in the work environment.

The NIOSH survey contained a number of questions related to aspects of safety climate, including management attitudes and actions related to safety, availability and use of personal protective equipment, job demands, worker involvement in safety, risk of occupational exposure to blood-borne pathogens, effectiveness of self-protective actions, and information and communications regarding UP and other safety matters. Using a 13-item scale to measure safety climate, respondents who perceived a strong commitment to safety at their institution were over two and a half times more likely to be fully compliant than respondents who did not perceive a strong safety climate (Gershon *et al.*, 1995). In a separate analysis of the nurses at the high HIV prevalence site, job hindrances were found to be the strongest predictor of compliance and safety climate was the best predictor of job hindrances (DeJoy *et al.*, 1995b). Safety performance feedback and availability of personal protective equipment were the strongest predictors of safety climate, together accounting for 30% of the variance.

9.8 SUMMARY AND CONCLUSIONS

The preceding discussion highlights several potential action alternatives for minimizing occupational exposure to blood-borne pathogens in medical care settings. The central issue is the need to pursue a broader, more multi-faceted intervention strategy that does not rely so heavily on the individual worker's ability to unfailingly follow safe work practices in all situations. The precautionary behaviour of individual HCWs is certainly important but significant opportunities exist to develop and apply engineering controls, to improve the design and organization of tasks, and to create organizations that facilitate and reinforce safe behaviour.

There is a definite need for careful analysis of specific patient-care tasks in the context of UP. This is an essential step in the development of engineering or passive control measures. Ideally, engineering controls provide workers with automatic protection but they can also be beneficial if they simply make it easier for workers to follow safe work practices. The job/task analysis shows that job-related hindrances play an important role in non-compliance. Some of the options for addressing this problem may reside within the worker analysis. In assessing the benefits and barriers associated with UP, HCWs may well include the benefits received by the patient when treatment is unencumbered by personal protective equipment. A possible way to alter this cost-benefit trade-off may be to enhance the self-efficacy expectancies of HCWs through skill- and strategies-based training. The objective here is to make HCWs more confident and comfortable in using protective equipment while providing care.

The environmental/organizational analysis underscores the importance of identifying and analyzing high-risk task environments. Compliance demands are seldom static and compliance is clearly more problematic in some situations than in others. Widely used job hazard analysis techniques should be applied to patient-care tasks and other potentially hazardous job activities. Macro-task analysis is also needed to fully examine the contribution of situational factors to non-compliance.

Finally, organizational safety climate is likely to be an important leverage point for improving UP-related behaviour. A positive safety climate is one in which a high priority is assigned to safety and where this commitment is demonstrated in both word and action. Three recommendations are offered in this regard. First, safety should be integrated into the management system of the organization. This means that safety should be handled in a fashion similar to other core functions of the organization. Second, a balanced view of accident/injury causation should be adopted; poor safety performance should not be viewed as simply a behavioural or worker-focused problem. To date, UP training efforts have focused almost exclusively on frontline HCWs. Supervisors and administrators should also be trained since they are essential to creating supportive safety climates. This training should emphasize the importance of organizational level action in achieving safety goals. Third, special emphasis should be placed on improving safety-related communication and performance feedback systems. An important step in this regard is to provide opportunities for two-way communication. UP-related findings (DeJoy *et al.*, 1995b) underscore the general importance of feedback to safety climate and suggest that some HCWs, most notably physicians, may often be 'outside the loop' in terms of regular safety communications and feedback.

Although this chapter has focused on HCWs and compliance with universal precautions, the work-systems approach is applicable to almost any occupational safety or health problem. The principal strength of this approach is that self-protective behaviour is examined in the context of specific job demands and broader organizational and environmental factors. However, even in situations where the self-protective actions of individual workers are less critical, important insights into prevention can be gained by examining worker–job/task–environmental/organizational linkages. Even engineering measures are effective to the extent that workers accept them, they do not compromise job performance and they are not undermined by decisions or actions at the organizational level.

In terms of broad-based occupational safety and health practice, the work-systems approach moves us in the direction of two important goals. First, the systems perspective underscores the importance of environmental and organizational factors in enabling and reinforcing safe workplace behaviour. This diminishes the tendency to examine worker behaviour removed from its job and organizational contexts, and helps move us beyond the 'blame and train' mentality. Second, the work-systems approach leads almost automatically to multi-faceted interventions and the use of complementary behavioural and environmental/organizational interventions. Comprehensiveness in designing intervention strategies is almost always superior to focusing exclusively on either the worker or the environment.

References

BECKER, M.H. (ed.) (1974) *The health belief model and personal health behavior*, Thorofare, NJ: Slack Press.

BECKER, M.H., JANZ, N.K., BAND, J., BARTLEY, J., SNYDER, M.B. and GAYNES, R.P. (1990) Non-compliance with universal precautions policy: why do physicians and nurses recap needles? *American Journal of Infection Control*, **18**, 232–239.

BOGNER, M.S. (1994) *Human error in medicine*, Hillsdale, NJ: Erlbaum.

CASEY, S. (1993) *Set phasers on stun, and other true tales of design, technology, and human error*, Santa Barbara, CA: Aegean.

CENTERS FOR DISEASE CONTROL (1987) Recommendations for prevention of HIV transmission in health-care settings. *Morbidity Mortality Weekly Report*, **36**(suppl. 2S), 1S–16S.

COX, T. and LEITER, M. (1992) The health of health care organizations. *Work and Stress*, **6**, 219–227.

DEJOY, D.M. (1996) Theoretical models of health behavior and workplace self-protective behavior. *Journal of Safety Research*, **27**, 61–72.

DEJOY, D.M., MURPHY, L.R. and GERSHON, R.M. (1995a) Safety climate in health care settings. In *Advances in industrial ergonomics and safety VII*, ed. A.C. Bittner, pp. 923–929, London: Taylor and Francis.

(1995b) The influence of employee, job/task, and organizational factors on adherence to universal precautions among nurses. *International Journal of Industrial Ergonomics*, **16**, 43–55.

GERSHON, R.R.M., VLAHOV, D., FELKNOR, S.A., VESLEY, D., JOHNSON, P.C., DELCLOS, G.L. and MURPHY, L.R. (1995) Compliance with universal precautions among health care workers at three regional hospitals. *American Journal of Infection Control*, **23**, 225–236.

HAMMOND, J.S., ECKES, J.M., GOMEZ, G.A. and CUNNINGHAM, D.N. (1990) HIV, trauma, and infection control: universal precautions are universally ignored. *Journal of Trauma*, **30**, 555–561.

HERSEY, J.C. and MARTIN, L.S. (1994) Use of infection control guidelines by workers in healthcare facilities to prevent occupational transmission of HBV and HIV: results from a national study. *Infection Control and Hospital Epidemiology*, **15**, 243–252.

HOFFMAN-TERRY, M., RHODES, L.V. and REED, J.F. (1992) Impact of human immunodeficiency virus on medical and surgical residents. *Archives of Internal Medicine*, **152**, 1788–1796.

JAGGER, J., HUNT, E.H., BRAND-ELNAGGAR, J. and PEARSON, R. (1988) Rates of needlestick injury caused by various devices in a university hospital. *New England Journal of Medicine*, **319**, 284–288.

JOB, R.F.S. (1988) Effective and ineffective use of fear in health promotion campaigns. *American Journal of Public Health*, **78**, 163–167.

KELEN, G.D., DIGIOVANNA, T.A., CELENTANO, D.D., KALAINOV, D., BISSON, L., JUNKINS, E., STEIN, A., LOFY, L., SCOTT, C.R.J., SIVERTSON, K.T. and QUINN, T.C. (1990) Adherence to universal (barrier) precautions during interventions on critically ill and injured emergency department patients. *Journal of AIDS*, **3**, 987–994.

KRASINSKI, K., LACOUTURE, R. and HOLZMAN, R.S. (1987) Effect of changing needle disposal systems on needle puncture injuries. *Infection Control*, **8**, 59–62.

LINN, L.S., KAHN, K.L. and LEAKE, B. (1990) Physicians' perceptions about increased glove-wearing in response to risk of HIV infection. *Infection Control and Hospital Epidemiology*, **11**, 248–254.

MCCORMICK, R.D., MEISCH, M.G., IRCINK, F.G. and MAKI, D.G. (1991) Epidemiology of hospital sharps injuries: a 14 year prospective study in the pre-AIDS and AIDS eras. *American Journal of Medicine*, **91**(suppl. 3B), 301S–307S.

NEW YORK STATE DEPARTMENT OF HEALTH (1992) *Report to the legislature: pilot study of needlestick prevention devices*, Albany, NY: New York State Department of Health.

OCCUPATIONAL SAFETY AND HEALTH ADMINISTRATION (1991) December 6, Occupational exposure to bloodborne pathogens: final rule. *Federal Register*, **56**, 64004–640182.

RIBNER, B.S., LANDRY, M.N., GHOLSON, G.L. and LINDEN, L.A. (1987) Impact of a rigid, puncture-resistant container upon needlestick injuries. *Infection Control*, **8**, 63–66.

RUBEN, F.L., NORDEN, C.W., ROCKWELL, K. and HRUSKA, E. (1983) Epidemiology of accidental needle puncture wounds in hospital workers. *American Journal of Medical Science*, **286**, 26–30.

SMYSER, M.S., BRYCE, J. and JOSEPH, J.G. (1990) AIDS-related knowledge, attitudes, and precautionary behaviors among emergency medical professionals. *Public Health Report*, **105**, 496–504.

WHITE, C.M. and BERGER, M.C. (1992) Using force field analysis to promote use of personal protective equipment. *Infection Control and Hospital Epidemiology*, **13**, 752–755.

YASSI, A. and McGILL, M. (1991) Determinants of blood and body fluids exposure in a large teaching hospital: hazards of the intermittent intravenous procedure. *American Journal of Infection Control*, **3**, 129–135.

YOUNGER, B., HUNT, E.H., ROBINSON, C. and McLEMORE, C. (1992) Impact of a shielded safety syringe in needlestick injuries among healthcare workers. *Infection Control and Hospital Epidemiology*, **13**, 349–353.

ZOHAR, D. (1980) Safety climate in industrial organizations: theoretical and applied implications. *Journal of Applied Psychology*, **65**, 96–102.

PART FOUR

Organizations, management, culture and safety

INTRODUCTION

If you pick up a book on safety now, in the late 1990s, you will find that a large part of the focus is on how the organization and work system influence safety. As discussed in the first chapter in this part, by Hale and Hovden, this was not always so. The focus on organizational factors and the management of safety is a relatively recent advance in our fight to make workplaces safer. It may now seem obvious that management and work systems play a major role in determining safety. This change in focus only occurred, however, when it was realized that the causes of accidents are not simple nor linear and that true prevention, that is the action needed to stop the accident or injury occurring at all, requires a much broader focus on what goes on in the workplace.

Research on the causes of all fatal accidents occurring at work in Australia over a three-year period demonstrated that when the wider circumstances of the accident occurrence are taken into account, factors occurring considerably earlier in time were a common feature of the majority of fatalities (Feyer and Williamson, 1991). In most cases these earlier factors were to do with work practices, or the way the work was organized and done, and as such were either formally or tacitly allowed to persist by company management. In addition, for each fatality, the events and factors which made a primary or root-cause contribution were discriminated for each fatality. A significant number of fatal accidents had multiple factors as root causes but again, in the vast majority of cases, the root cause was a pre-existing feature of the work situation and not a part of the series of events leading immediately to the fatal accident. Again the most common root causes were pre-existing work practices, that is the way the work was normally done. This research demonstrates, therefore, from a population-based data set, that not only do work organizational or management factors play a role in workplace fatalities, their role is pivotal.

Building on this empirically derived evidence, Hale and Hovden, in their chapter in this part, identified four additional factors which played a timely role in stimulating interest in the role of organizational factors and management in safety. This meant that, as a result, in the search for the causes of the significant number of major disasters which occurred at the end of the 1970s and early 1980s, organizational factors were included in the search for contributing factors. It also meant that organizational and human factors started to be included in the tools used to estimate risk in workplaces, such as probabilistic risk assessment and considerably later the development of methods for measuring safety climate. All of these developments led to the realization that targeting of strategies for prevention of fatal and less severe workplace accidents must start with changing the way the work is done and organized.

In his chapter in this part, Wagenaar takes a similar view in his discussion of the origins of individual human failures. He argues that these errors or substandard behaviour are triggered by environmental factors. In Wagenaar's view the environment provides the reasons for action so that behaviours become errors because of the context in which they occur. As such these substandard behaviours are created more by management decisions and actions than by individual factors like motivation or attitude. The obvious consequence for accident prevention, Wagenaar argues, is not to make the individual worker responsible for avoiding substandard behaviours or to attempt to use more protective barriers, not to prevent the poor behaviour occurring but to reduce its consequences. He argues that the solution lies in trying to change the environments that promote the substandard behaviour.

The main subject of the chapter by Hale and Hovden is an in-depth look at what is currently known about the role of management and organizational culture in causing workplace injury and accidents. The authors take a broad look at the different approaches to the issue of organizational and management contributions to safety. Their review covers aspects from the development of management-related safety indicators to the sociotechnical and sociological theories of the role of organizational factors in safety. This review provides a summary of the relative contributions of the 34 identified approaches to organizational factors and safety. As such, the review serves a particularly useful purpose in tying together the very disparate literature on organizational factors and also in identifying the problems that beset the area.

The search for the major causes of occupational accidents has really moved now to concentrate on the systemic and organizational aspects of work rather than on the immediate circumstances and behaviours that lead up to each individual accident. This has, as a result, significantly changed the prescribed actions needed to reduce occupational accidents. Prevention has now pivotted in the direction of organizational indicators like the presence of an active safety committee, of safety rules and of the culture of safety in the organization. It is important, however, that the move to organizational and management approaches does not become too single-minded. We must be careful that we do not become too focused on these factors as the only ones of concern. Not all injuries are caused by management failures, even though it seems very likely that many of them are. The lesson that we need to take from this increasing move to be concerned with organizational aspects of accident causation is that these interact with other events and factors occurring later in time to cause the accident. Good accident prevention requires a consideration of all aspects of the system, including how individual workers interact with it.

References

FEYER, A.-M. and WILLIAMSON, A.M. (1991) A classification system for causes of occupational accidents for use in preventive strategies. *Scandinavian Journal of Work, Environment and Health*, **17**, 302–311.

FEYER, A.-M., WILLIAMSON, A.M. and CAIRNS, D.R. (1997) The involvement of human behaviour in occupational accidents: errors in context. *Safety Science*, **25**, 55–65.

CHAPTER TEN

People make accidents but organizations cause them

WILLEM A. WAGENAAR

10.1 INTRODUCTION

There are two assumptions that underlie the psychologist's interest in accident prevention: that people cause accidents and that their behaviour can be changed.

In his career as an accident analyst the author has looked at hundreds of accidents, from very small, like the cook on an offshore platform cutting his thumb when peeling potatoes, to the loss of entire platforms. The area of interest ranged from police shooting accidents, patients dying in intensive care wards, submarine and other shipping accidents, to accidents in chemical plants, energy production plants, refineries and the entire range of oil exploration and production. None of the accidents occurred without the essential contribution of human behaviour; which supports the first assumption.

However, why the second assumption? Why do we believe that human behaviour can be changed? The author, as a psychologist, believes strongly that human behaviour is lawful and therefore predictable; not 100% but to such a large extent that it is useful for accident prevention. The concept of lawfulness and of predictability runs counter to the intuitions that people cherish about themselves. Many people feel that they have a free will, that they differ from others, that their behaviour is based on intelligent and highly individualized considerations, and therefore that their behaviour is not easily predicted and not subject to simple laws. This can be called the *illusion* of unlawfulness. This is not to deny free will or individuality but these factors have only a small contribution to our everyday behaviour. Most of it, say 99%, is not based on free choice and is not an expression of our individual personality. It consists of highly automatized reactions to environmental conditions. You, as a reader, are reading this text from the top of the page downwards. You are reading the lines from left to right. You have never seen this text before but you have developed an automated reaction to printed text, as much as the author has developed an automated writing habit. The generalization of writing and reading habits is quite useful because it enables us to communicate without knowing one another and without coordination of our free wills. Free will and personal choice would only disturb the process.

10.2 SKILL-BASED AND RULE-BASED OPERATION

Reading and writing are examples of what Rasmussen (1982, 1983) has called *skill-based* operation. Skill-based behaviour evolves as a series of steps in a well-rehearsed routine or *schema*. Once the schema is activated, possibly as a result of a wilful decision, the steps follow in a prearranged order. For instance, if I decide to go home after a day's work, the schema for going home is started and followed automatically. I get my things, leave my office, find my car, enter it, start the engine, get on the motorway, follow a well-known route, etc. Each of these steps contain a number of substeps. Entering my car consists of finding my keys (usually in the right-hand pocket), identifying the car key (with the black plastic cover), inserting it into the lock (teeth upwards), turning it (clockwise), waiting for the click, removing the key, pulling the door handle. Probably I have left out a number of steps but I rarely leave them out when I actually perform the task. This is typical for skill-based or automated behaviour: one does it without explicitly knowing how it is done. This is possible because the end of step x is the stimulus for the initiation of step $x + 1$. Seeing the lock in the door of my car tells me to insert the key and how to insert it. The sight and the feeling of the inserted key tell me to turn it and in which direction. The click tells me to withdraw the key, etc. Free will and personal choice are not useful in these stages.

   Now consider what went wrong on some days when I went home in the last few weeks:

- I forgot to take home my students' tests, which I intended to score over the weekend;
- I did not remember that my car was parked in a different place;
- I did not realize that this morning I had come by train, not by car;
- I forgot to buy gas, although this morning I had noticed that the gas tank was empty;
- I drove home, although I had agreed to meet my wife in the theatre.
- I left my keys on the table in my office; I had used them when I went to the archive room without my jacket on and had not put them back in the usual place; and
- I entered the car key upside down when I was distracted by the sight of another car making a dangerous manoeuvre on the parking lot.

   The first six of these problems can be characterized as a failure to abandon the fixed schema. Usually I do not take tests home. Usually I do not have to buy gas. Usually I do not have to look for my keys when leaving the office because they are in my pocket. The last problem is characterized as a disturbance in the execution of the schema. The omission of actions that are not parts of fixed schemas is an example of too little personal choice, rather than of the opposite.

   There are many ways in which skill-based operations can go wrong (Reason, 1990). The schema can be triggered at the wrong moment or not at all. One can switch from one schema to another because the two have some elements in common. The schema can be abandoned because activation drops. Most human errors stem from such disturbances in skill-based operations. We call them *wrong but weak* because they are usually detected and corrected in time. I do not go home without my car keys because I simply cannot but I may arrive at home without the tests because nothing is reminding me to take them with me. Hence, although errors in skill-based behaviour are weak, they may frequently persist because the occasions for making such errors have a high base rate. It is interesting to note that out of 200 errors in nuclear power plants, listed by Rasmussen (1980), over 60 were classified as 'forgetting to do something', which was either following a well-known schema when something else was required or not following a skill-based schema because of some sort of distraction or disturbance.

When the skill-based schema is stopped, for instance because the car keys are not in my pocket, there obviously is a problem that needs to be solved. The first attempt is to solve it through *rule-based operation*. The problem is identified as belonging to a familiar class of problems, for which there is a well-known solution. When my car keys are not in my right-hand pocket, the solution is:

- first try all other pockets;
- then try the right-hand pocket again;
- try all other pockets again;
- repeat these steps frantically; and
- return to my previous location and look there.

Again, there is little need for creativity or free will at this level. The classification of the problem as 'look for something lost' is automatic and the solution comes automatically with the classification.

10.3 THE GENERAL ACCIDENT CAUSATION SCHEME

Normal occupational life consists almost entirely of two levels: *skill-based* and *rule-based* behaviour. It is governed by well-rehearsed schemata and problem–solution routines, and triggered by environmental stimuli. This is my reason for suggesting that human behaviour is lawful and predictable, as far as we are concerned with the prevention of occupational accidents. The general accident causation scheme, shown in Figures 10.1 and 10.2, is based on this idea of automaticity. The first part, called the proximal section because in time and location it is close to the actual accident, shows a frequently encountered sequence: *substandard actions* cause *disturbances*; if there are no *barriers* against these disturbances, or if the barriers of safety devices are breached, we will have an accident. Traditionally attempts are made to break this pattern either by having more barriers or by telling people to omit substandard actions. There is a limit to both approaches. Increasing the number of safety devices may help up to a certain point; beyond that point it may be counter-productive. Irrespective of how many safety devices are put in place, the system will never be foolproof. Telling people not to commit substandard actions does not take into account that most errors stem from disturbances in skill-based behaviour, which by definition is not under conscious control. If such an advice is interpreted as omitting all skill-based behaviour, then it is even more unrealistic: the

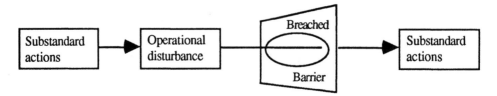

Figure 10.1   The proximal section of the general accident causation scheme.

large majority of professional tasks can *only* be performed at the skill-based level. The ensuing disturbances are a logical consequence and cannot be avoided by telling people to act safely. The attempt is rather like Kletz's (1985) famous example of an instruction to pilots: 'Do not crash this plane'.

It is only realistic to assume that the system of safety barriers is not foolproof and that people cannot be simply told to act safely. How then can we still influence behaviour so that accidents are avoided? The answer is in looking beyond the proximal part of the accident causation scheme. Substandard actions, operational disturbances and breached barriers occur at the shopfloor, close to the actual location of the accident. However, the substandard behaviour is usually skill-based or rule-based, which means that it follows known patterns and that it is triggered by environmental factors following familiar laws of behaviour. The easiest way of understanding this is by realizing that the environment provides the reasons for doing things, even when they are wrong, dangerous and forbidden by established rules and procedures. Violation of rules in a maintenance job is not a random phenomenon but is causally related to such factors as time pressure, the impractibility of the complete procedure, unawareness of the correct procedure or the physical impossibility to follow the prescribed procedure. The course of events preceding substandard actions is sketched in the *distal* section of the general accident causation scheme, presented in Figure 10.2. It is called 'distal' because these events usually take place far away from the actual accident, with respect to both time and location.

Environmental conditions that elicit substandard acts are called *latent errors*. They are latent because they cannot be recognized directly as errors. Usually they are present without causing immediate threats. People have learned to cope with them, maybe through the adoption of substandard behaviour, without causing accidents. They are called errors because they have the potential of being a sufficient step towards an accident. There is an infinite number of latent errors but they can be classified into 11 general failure types (GFTs):

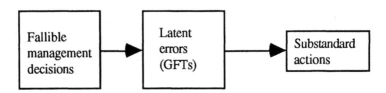

Figure 10.2   The distal section of the general accident causation scheme.

- hardware (HW)
- design (DE)
- maintenance (MA)
- procedures (PR)
- error-enforcing conditions (EE)
- housekeeping (HK)
- incompatible goals (IG)
- communication (CO)
- organization (OR)
- training (TR)
- defences (DF)

A full explanation of these categories is given in Wagenaar *et al.* (1994). What matters here is that these GFTs are descriptive of work situations, not of people. They are not like 'awareness', 'attitude', 'personality', 'motivation' or 'safety culture'. The GFT training category means that there is no appropriate organization to ensure the proper training for the jobs that people do. The reason for this emphasis on environmental factors is that skill-based and rule-based behaviours are only to a small extent affected by person-related factors. Situational factors like equipment, procedures and planning are far more influential. Moreover, as shown in Figure 10.2, GFTs are created by management decisions and therefore are much more under the control of management than person factors such as attitude or motivation.

10.4 THE REMOVAL OF UNWANTED BEHAVIOUR

The general belief that psychologists can cost-effectively change person-related factors, such as attitude or personality, and indirectly the behaviour that follows from them, even when situational factors invite the wrong behaviours, is totally unfounded. In most situations it is more practical to remove the latent failures that elicit the unwanted behaviour. The problem, however, is that the removal of latent failures, or generally the control of GFTs, is in the hands of management. Decisions are needed that ensure the proper design of workplaces, effective maintenance of installations, up-to-date formulation of work procedures, careful planning of operations, effective communication among all levels of the organization and a realistic selection of company goals. If the company wants to make an extra profit at the cost of time pressure or of running increased risks, there is nothing that operators at the shopfloor can do about it, apart from refusing to do their jobs. Even when all the signals of approaching doom are visible, such as frequent breakdown of equipment, habitual violation of rules, increased frequency of potentially damaging incidents, sickness and absenteism, and loss of communication between management and the workforce, there is still nothing that can be done to increase safety at the work floor, unless management wants it, originates it and keeps supporting it. Hence management has a dual task in the management of safety: to monitor the occurrence of latent failures in the 11 GFT categories and to take action when one or more GFTs become too dangerous.

10.5 THE TRIPOD INSTRUMENTS

Our groups in Leiden and Manchester have developped a number of techniques for the monitoring of GFTs. They can be divided into *reactive* and *proactive* techniques. A reactive technique is used as a reaction to a specific incident or accident, and helps to identify the causal factors that contributed to the occurrence of the unwanted event. A proactive technique probes the state of the 'safety health' of an organization without using incident or accident material. Both types of technique produce an overview of latent failures in an organization, summarily presented in a safety state profile, as presented in Figures 10.3 and 10.4.

Both types of analysis use questionnaires for the collection of data. The reactive analysis employs a questionnaire that analyses the structural background of events that caused the incident or accident. The most reliable instrument of this type has been developed by Van der Schrier (in prep.). Coefficients of validity are in the range of 0.70 to 0.90. The instrument can be used by people who have an expertise with respect to the particular operation; no specific expertise in accident analysis is needed. The proactive analysis employs questionnaires that take account of symptoms of unsafe operation and relate these symptoms to their structural background. The application of such questionnaires is more time-consuming than in the case of reactive analysis because the proactive questionnaires must be specific for the type of operation. A number of questionnaires are available now, especially for application in the oil industry. Again there is no special need for users to be experts in the area of safety analysis. Both reliability and validity were demonstrated to be in the order of 0.70 to 0.90 for these questionnaires (Hudson *et al.*, 1994). Figure 10.4 illustrates that two parallel versions of the questionnaire yielded more or less the

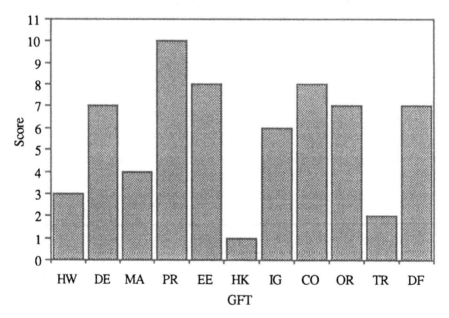

Figure 10.3 Safety state profile representing the reactive analysis of a gas incident on an offshore production platform.

same result. The collective technique in these instruments, based on the notion of reactive and proactive analysis of GFTs, is called TRIPOD.

The purpose of the TRIPOD instruments is to show management what *they* must do in order to reduce or remove GFTs that otherwise will elicit substandard behaviour. The indications to management can be quite specific, such as to issue procedures in more languages or to test the skills of operators after they have participated in a training course. The critical question, of course, is whether a company will follow such indications. It is my experience that the major obstacle is a management's belief that they can live with a high level of GFTs and still improve the behaviour of employees by instruction, threat, reward, attitude improvement, i.e. measures that affect only the proximal section of the accident causation scheme. This belief ignores the simple observation that the relationship between latent failures and the subsequent substandard behaviour is almost automatic because it is controlled by skill-based or rule-based operation. The environmental factors that control skill-based and rule-based behaviour can only be changed by measures in the distal section of the accident causation scheme.

If managers cannot understand the logic of the TRIPOD rationale they may hope that they can limit the measures intended to improve safety to cheap changes at the shopfloor. However, this would ignore a mass of psychological evidence,

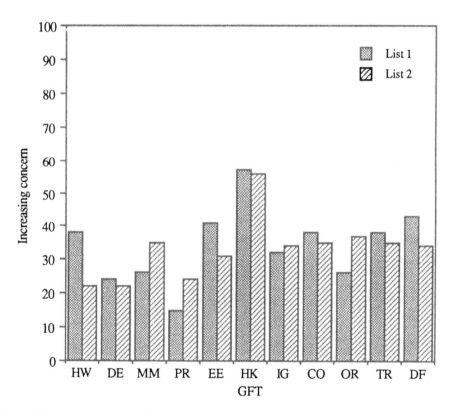

Figure 10.4 Safety state profile representing the proactive analysis of safety on 30 ships of the Dutch merchant marine. List 1 and 2 refer to parallel versions of the proactive questionnaires.

which is highly relevant because safety is a topic that belongs in applied psychology: it is about how organizations can control the behaviour of people who may cause accidents.

References

HUDSON, P.T.W., REASON, J.T., WAGENAAR, W.A., BENTLEY, P.D., PRIMROSE, M. and VISSER, J.P. (1994) TRIPOD delta: proactive approach to enhanced safety. *Journal of Petroleum Technology*, **46**, 58–62.

KLETZ, T.A. (1985) *An engineer's view of human error*, Rugby: The Institution of Chemical Engineers.

RASMUSSEN, J. (1980) What can be learned from human error reports? In *Changes in working life*, eds K. Duncan, M. Gruneberg and D. Wallis, London: Wiley.

(1982) Human errors: a taxonomy for describing human malfunction in industrial installations. *Journal of Occupational Accidents*, **4**, 311–335.

(1983) Skills, rules, knowledge: signals, signs and symbols and other distinctions in human performance models. *IEEE Transactions: Systems, Man and Cybernetics, SMC-13*, 257–267.

REASON, J.T. (1990) *Human error*, New York: Cambridge University Press.

VAN DER SCHRIER, J.H. (in prep.) Reactive tripod: the development of a tool for accident analysis, Leiden University (thesis in preparation).

WAGENAAR, W.A., GROENEWEG, J., HUDSON, P.T.W. and REASON, J.T. (1994) Promoting safety in the oil industry. *Ergonomics*, **37**, 1999–2013.

CHAPTER ELEVEN

Management and culture: the third age of safety. A review of approaches to organizational aspects of safety, health and environment

ANDREW R. HALE AND JAN HOVDEN

11.1 INTRODUCTION

Attention to the management of safety, health and environment is as old as civilization. The Babylonian king Hammurabi incorporated a provision in his code of laws in the second millenium BC ordering the punishment of the mason if the house he built fell down and killed the owner. The punishment might seem excessive now (death of the mason if the owner died and death of his son if the owner's son died) but the principle that a company is liable for managing its affairs so that it produces a safe product in a safe way is surprisingly modern. This attention to management was, however, largely concerned with government regulation requiring managers to make technical and human resource provisions in the name of safety. It did not go into how the company should manage that provision. Attention to prevention was practical rather than scientific.

The first age of more scientific study of safety concerned itself with the technical measures to guard machinery, stop explosions and prevent structures collapsing. It lasted from the nineteenth century through until after the Second World War and is characterized by such statements as those made by UK factory inspectors in the late nineteenth century that the only accidents they were interested in having reported were those with technical causes, since others could not be reasonably prevented (see Hale (1978) for a discussion of this interpretation of the scope of 'safety management').

The period between the world wars saw the development of research into personnel selection, training and motivation as prevention measures, often based on theories of accident proneness (see Hale and Glendon (1987) for a review). This ushered in the second age of safety, which led a rather separate life from the technical measures until the period of the 1960s and 1970s, when developments in probabilistic risk analysis and the rise and influence of ergonomics led to a merger of the two approaches in health and safety. The exclusive dominance of the technical view of safety in risk analysis and prevention was broken and the study of human error and human recovery or prevention came into its own. This period began with slender reviews of the available material (CIS, 1967; Surry, 1969; Hale and Hale, 1972). The achievements of that period are documented in the increasing number of large standard texts reviewing and developing the human factors literature (Hale and Glendon, 1987; Petersen, 1989; Reason, 1990; Glendon and McKenna, 1995). Already the field has expanded beyond the possibility of confining it in its full details within the covers of one book (see, for example, Kirwan (1994) on human reliability assessment techniques and ReVelle (1980) on safety training).

Just as the second age of human factors was ushered in by increasing realizations that technical risk assessment and prevention measures could not solve all problems, so were the 1980s characterized by an increasing dissatisfaction with the idea that health and safety could be captured simply by matching the individual to technology. The 1990s are already well into the third age of safety, where management systems are the focus of development and research. Just as the human factors age had a scientific literature dating back to the 1920s in the form of studies of accident proneness and motivation, so the field of safety management has a history of texts, usually traced back to the first edition of Heinrich's ground-breaking safety management textbook in 1931 but also traceable to the Safety First movement in the UK in the 1920s and the development of the National Safety Council in the USA (Dwyer, 1991). That literature can be characterized, at least until the 1980s, as accumulated common sense and as general management principles applied to the specific field of safety. Heinrich's book in its successive reprintings was based on the analysis of many thousands of accidents but the management principles which he expounded were largely unsupported by specific field research of success or failure of specific management practices in relation to safety performance. The auditing approaches pioneered by Bird (Bird and Lofthus, 1976), based on Heinrich's principles, were also codified collections of practical expertise and judgement from people with long years of work in industry. They had, and still have, little or no scientific basis. Safety management texts, in English at least, were predominantly American in origin and UK industry turned explicitly to the US to learn about such activities as auditing.

The sociotechnical management literature (Trist and Bamforth, 1951; Thorsrud, 1981; Elden, 1983) did have a profound effect in Scandinavia in shaping the legislation towards the working environment. Through the studies undertaken by the Robens Committee in the UK in preparation for the 1974 Health and Safety at Work Act and through comparable influences in The Netherlands, that influence did trickle into the formulation of the principles of self-regulation underlying regulatory reform in those countries. However, it did not give rise to a separate literature on health and safety management. The first explicit applications to safety in the English language literature date only from the early 1980s (e.g. Robinson, 1982). The social organizational theory of Lewin (1951) did have an influence on the introduction of participative management in safety (e.g. Simard, 1995).

As late as 1992 a comprehensive review of the field of risk by the Royal Society (1992) in the UK, although it had a chapter on risk management, had no review of research or practice at the level of company management. Risk management was seen entirely at the level of society and regulatory action, apparently operating directly on the level of 'risk perception', the topic of the report's preceding chapter. The report *Organising safety* by the Human Factors study group of the Advisory Committee on Safety in the Nuclear Industry (ACSNI, 1993) appears to be the earliest UK attempt to review research in the field comprehensively (reviewing 147 references). In the US the studies initiated by the Nuclear Regulatory Commission after the Three Mile Island accident also carried out an extensive review (Haber *et al.* (1990) report that their assessment instrument was based on reviewing 3500 references to safety and organization). The summary of the conclusions of the World Bank workshops in 1988 and 1989 (Rasmussen and Batstone, 1991) also provides a partial review of work, notably in the high technology industries.

Research into the management of safety has increased enormously in the last ten years. A recent literature search by the authors in preparation for a textbook on the subject resulted in identifying over 1000 significant references to books, papers and reports on the topic in the scientific literature in English, French, Dutch and Scandinavian. This chapter is not an attempt to review all of that literature in detail nor to duplicate the review *Organising safety*. Its aim is to survey the field at a global level, to indicate the approaches taken by the principal groups of researchers, the reasons lying behind their studies and the organizational theories and models they have used. The chapter starts by looking briefly at the driving forces that have focused interest on safety management and derives a broad classification of studies from that analysis. The approaches taken in specifying or evaluating organizational factors are then classified. A tabulation of the topics that are emphasized is provided and these are discussed within the four frames proposed by Bolman and Deal (1984) for analyzing organizations. We draw conclusions about the trends and gaps in knowledge such a survey reveals. The chapter concludes by sketching a research agenda.

11.2 SCOPE AND DEFINITIONS

The aim of this chapter is to cover studies of the management of safety, health and environment, since these are increasingly named in one breath as inseparable objectives. We should state at this point, however, that by far the dominant one of these three in this chapter is safety. Research into environmental management at the level of the organization appears, as yet, to be very limited. The literature has concentrated on the development of certifiable management systems based on quality management principles (British Standards, 1992), on audit tools or on direct assessments of the costs and benefits of environmental technology (e.g. Huisingh, 1989). Studies are only recently appearing on environmental culture (e.g. Wehrmeyer, 1995). The concentration of research in the environmental field has been on the level of technology assessment and supra-organizational studies of technology and product chains and on regulatory tools for encouraging sustainable development.

Research into health management at company level has also remained largely at the level of technology and individual behaviour, or has been subsumed as an element in the safety management system (e.g. Det Norske Veritas, 1994). The one exception to this is the major emphasis on some aspects of psychological health in the studies of participative management and the quality of working life.

Where our intention is to refer to the whole area of safety, health and environment we will use the term 'SHE management'. If only one of the three words is used, it implies the narrow context.

This chapter concentrates on the literature dealing with the internal management system of organizations. We have not attempted to review the vast literature on risk management at the national or industry level dealing with regulation, standard setting, risk policy, enforcement, etc. The boundary with this literature is the limited number of regulatory codes and standards specifying what the management structure or functioning of companies should be (e.g. HSE, 1985; BS, 8800). We have also set a lower boundary that excludes the management of individual workplaces. This is a somewhat arbitrary distinction since a small company may be no bigger than a single work group in a larger company. Most of the studies we review have concentrated largely on the overall management of SHE, encompassing the whole of the life cycle of the organization. There are, however, a few research traditions and approaches which have had a significant influence on management studies, despite having concentrated on the work group. These concern notably participative management studies and studies of high reliability organizations, which concern themselves with 'on-line' management of risk, as opposed to the 'off-line' concern with management structure found in much of the literature. At times the line is hard to draw and may seem arbitrarily to exclude valuable literature. Thus we have not included as a separate line of work the Scandinavian studies of safety analysis and information systems in companies or of feedback of safety performance as a method of safety improvement (e.g. Kjellen and Larsson, 1981; Kjellen and Baneryd, 1983; Saari and Näsänen, 1989; Carter and Menckel, 1990) since they deal only with parts of the management system. A number of these Scandinavian studies do, however, articulate with the participative management tradition, which is dealt with below.

Finally we have confined our review to literature that has some scientific or scholarly basis rooted in the study of SHE issues. This excludes two sorts of text: the general management and organizational literature, which usually mentions SHE only in passing and treats its control as an extension of general management principles, and the largely anecdotal literature of many managers, safety specialists and consultants giving descriptions of how they have managed safety or experienced safety management.

11.3 SPRINGS OF ACTION

The interest in research and action on organizational factors and management systems blossomed in the late 1980s. We identify four interrelated factors in this development, which have led to six dominant motivations for research.

11.3.1 Analysis and prevention of major disasters

Many of the research studies on organizational factors trace their origin to, or perhaps more accurately use as a touchstone, a handful of major disasters in the nuclear, petrochemical and transport industries. Time and again the disasters at Seveso, Bhopal, Chernobyl and Piper Alpha, the crash of the Challenger, the sinking of the *Herald of Free Enterprise*, the fire on the *Scandinavian Star*, the rail accident at Clapham Junction and the King's Cross Underground fire are cited. While the organizational factors in these seminal disasters are clear, it is also plain that many disasters before provided just as many organizational lessons to be learned (see Bignell *et al.*, 1977; Turner, 1978; Booth, 1980; Perrow, 1984; Fortune and Peters, 1995). The official and dominant scientific investigations of the earlier disasters did not, however, focus on these, but on the technological and human factors. It would appear that the time was ripe in the late 1980s for the critical examination of management, rather than that management only began to fail seriously at that time. This may have been due to the fact that the disasters took place in high technology industries with well-developed, often highly bureaucratic, safety systems, which had been thought until then to be proof against such major disasters. The reports of the disasters themselves therefore looked beyond such structural systems to try to find the reasons why they did not work (see e.g. Department of Transport, 1987, 1988, 1989; Department of Energy, 1990). These high-profile 'management accidents' are referred to so prevasively in the literature that they cannot have been said to have given rise to any one sort of research. As such we do not identify this as a separate line of research purpose in Table 11.1.

11.3.2 Modifying probabilistic risk assessment (PRA)

One of the regulatory responses to major disasters was the introduction of mandatory quantified risk assessment in the nuclear industry, followed in some countries by the petrochemical industry. Early PRAs were almost exclusively technical. The pioneering work of Rasmussen and the WASH-1400 report can be seen as the beginning of this work. After Three Mile Island the human factor was added, after a great deal of intensive work on developing methodologies (see e.g. Kirwan, 1994). With the introduction of risk acceptance criteria, sometimes linked to land-use planning as in The Netherlands, where quantitative risk contours around major hazard plants were required to be defined, companies began to complain that the use of generic failure rates in the calculation of the risk contours did not give due allowance for the good safety management systems they had in place. They pressed to be allowed to reduce predicted probabilities if they could prove that they had a good management system. From the other side regulators, influenced by the major disasters mentioned, pressed for incorporation into PRAs of the effects of the common cause failures introduced by underlying poor safety management factors, in order to increase risk estimates. The common feature for both sides was the need to develop methods of assessing management systems. This purpose is labelled 1 in Table 11.1.

11.3.3 Deregulation and self-regulation

A combination of cut-backs in government activity under right-wing regimes in countries such as the USA, the UK and Australia, the threatened bankruptcy of social insurance systems in others (e.g. Sweden, New Zealand, The Netherlands) and a reappraisal of the success of traditional detailed legislation (e.g. in the UK and Norway) has led to the withdrawal of government from detailed technical regulation and close shopfloor inspection of health and safety. Under the banner of self-regulation (e.g. Robens, 1972; Norwegian Ministry of Local Government and Labour, 1987) the emphasis has been changed to place the central responsibility on company management for devising, installing and monitoring their own safety management system. This has produced two stimuli for studies of organizational factors.

High-level indicators for regulatory attention and control

Regulatory authorities have become interested in indicators that can provide them with a basis for planning the more limited controls they still wish to undertake to replace the more detailed worksite and plant inspections. They have sought these indicators at the level of the management system, its adequacy and functioning. The interest here is in early warning signs to focus the scarce regulatory resources (e.g. Shrivastava, 1992) or periodic assessments to give a certificate that the company system is capable of self-regulating. This research purpose is labelled 2 in Table 11.1.

Indicators for detailed monitoring and management control

Company managements that have taken seriously their responsibility to set up self-regulating systems have increasingly realized the need for extensive and detailed monitoring systems linked to performance indicators throughout their activities, in order to motivate and guide employees (in particular line managers) to manage safety, health and environment actively and consciously. Traditionally, audits based on accumulated experience have filled this need but some companies have felt the need to provide a more scientific basis for these tools. Some regulators (e.g. Lindsay, 1992) have supported such developments, partly with another motivation in mind, namely to counter the widespread tendency of managers to place the blame for accidents on the injured worker. Studies of the underlying, or 'root', causes of accidents have been undertaken to point out the importance of management control decisions. Studies arising from this tradition have tended to consider a much more detailed and specific level of management factor than those in the previous section. Research in this area shades into the level of group dynamics, when it concerns itself with the performance of on-line risk management, e.g. in control rooms.

This tradition has also been fed from the concern in high-technology industry for quality management and for mission reliability. The US defence and aerospace industries codified much of the practical wisdom in this area in the MORT manuals (Johnson, 1980). Later concerns with standards at international level for quality management (ISO, 1987) have merged with this tradition and codify it in terms of basic principles of management control and feedback. This has been the dominant strand in safety management for many years and is identified as purpose 3 in Table 11.1.

An offspring from, or reaction to, this line has been the work that has tried to get behind the structural view of management and look at the 'motor' for performance in SHE, which has been labelled 'safety climate' or sometimes 'safety culture' (but see below for our criticism of this confusion of terms). This work has still tried to enumerate the factors of successful management but it has been a search for its subtler aspects not found in policy statements and safety manuals. It has been partially informed by the organizational culture literature (e.g. Hofstede, 1986), which has its roots in the differences found between companies in the same technology operating in different countries. Research in this line is identified as purpose 4 in Table 11.1.

11.3.4 Developmental studies

More recently there has been an interest in the dynamics of management systems. This has its roots in the studies of the Tavistock Institute, which were applied in the early years mainly to quality of working life and job satisfaction. Only later did they come to be applied also to safety management. As more companies have started to undertake the development of systems, both they and the regulatory authorities have become interested in the way in which systems can be implemented, how companies can learn to improve on existing systems and whether a combination of management systems for health and safety, environment and quality is feasible, desirable or essential. Such studies have focused not so much on the static management system but the motivation for, and process of, organizational learning. In this respect they also pick up the theme of 'continuous improvement' from the studies of quality management. This line is labelled 5 in Table 11.1 under 'purpose'.

A characteristic of this tradition is a concern with matching the management system to the type of organization, based on an organizational classification (especially that of Mintzberg, 1980). One of the conclusions from our review is that this line has not yet been well developed but we identify it separately in Table 11.1 as purpose 6.

11.4 STUDIES COVERED

Despite the restrictions described in the previous section, the literature on safety management is now very large and one chapter cannot pretend to review it fully and comprehensively. We have therefore been selective both in the material reviewed and referenced here, and in the type of conclusions we try to draw. We can only apologize to readers who feel that we have missed an important approach or set of papers and challenge them to fill the gap with their own review.

In reviewing such a vast literature it is difficult to arrive at a convenient means to reduce the complexity to a manageable extent without losing all of its richness. We have chosen to group the papers into 'approaches' or 'lines' of research. By this we mean a defineable line of research conducted either by one research group (identified in the list below and in Table 11.1 by the name of the university, research institution or [type of] consultant from which the team operates) or by a community of researchers from many different research groups or consultancies applying a recognizable research paradigm or management approach and referring intensively to each other's papers or management tools (these are identified by the school or discipline from which they come or by the type of research methodology they use). Sometimes we have to make an arbitrary choice of one of a number of researchers or commercial organizations since there are too many to mention. We have then tried to select a 'typical' representative or one who has been particularly explicit about their contribution or theory. This choice is irreducibly personal but we hope it does not bias the conclusions too much. This process has produced 34 approaches or research lines, which are summarized below and analyzed in Table 11.1. A brief description of each line is given first, followed by a description of the characteristics of their approach that are contained in the table. In the bibliography at the end of the chapter some representative references to papers by each group are given together under the name and study number used in the table. The order in which the approaches are given follows our classification of them according to the six dominant research purpose derived from the earlier discussion:

1. Incorporation of a management factor into a PRA.
2. Discovery of high-level management/performance indicators for regulatory guidance.
3. Indicators of the structure and functioning of successful SHE management systems.
4. Indicators of SHE climate, including studies of the SHE implications of different organizational cultures.
5. Studies of the development and integration of SHE management systems.
6. Studies matching organization type explicitly to SHE management system factors.

11.4.1 Lines of research identified

The following is the list of research lines, identified either by research establishment or research approach. It gives a brief description of the type of approach taken and complements the material in Table 11.1.

1. *UCLA.* Analysis of major disasters and development of the work process assessment method to incorporate organizational factors into probabilistic safety assessments via their influence on task and work process completion.
2. *Birmingham.* Development of a method based on human reliability assessment (HRA) techniques to incorporate safety culture failure mechanisms in PRAs.
3. *EDF (Eléctricité de France).* Critique of PRA for failure to incorporate organizational factors. Development of a safety climate scale, accident modelling techniques and occupational stress assessment tools as supplements.
4. *Technica/Four Elements.* Development of management factor weighting for PRA based on an audit of control and monitoring loops for eight main aspects of management of loss of containment derived from incident analyses of pipe and vessel failures.
5. *Science Applications/Maryland.* Work funded by the American Nuclear Regulatory Commission (NRC) to establish safety indicators. Development of six levels of modelling of influences, linked to triads of risk maker, manager and assessor. Modelling of influence of factors at organization and policy levels.
6. *Minnesota.* NRC-funded work to develop high-level safety indicators. Correlational studies of indicators at outcome, intermediate and antecedent levels, linked to safety performance measures and company violations.

7. *NIOSH et al.* Studies initiated in the USA by NIOSH in the 1970s to compare matched pairs of high and low accident companies. A scattering of later studies using the same paradigm is available in the literature.

8. *Imperial College.* Case studies of the safety management systems in companies in the aftermath of the introduction of framework legislation in UK. An assessment of the success of, and limits to, self-regulation.

9. *Commercial auditors.* Commercially available audits or audits developed individually by major European and international companies are widely used. Many audit systems are not open to scientific study for reasons of commercial secrecy. For the purpose of this chapter two representative audits with a leading place in the international market were used.

10. *Health and Safety Executive.* Studies carried out by the Accident Prevention Advisory Unit, with publication of guidance for industry.

11. *Government (Dutch).* The official guidance notes of the Dutch Ministry of Social Affairs and Employment for implementation of a working conditions management system are taken as an example of a second national set of guidelines.

12. *US safety management texts.* This category is also not exhaustively reviewed. Two highly influential authors of American texts (Heinrich and Petersen) are taken as illustrative of this tradition.

13. *Chemical process industry.* This group is represented by the development of audit systems by the American Centre for Chemical Process Safety based on the OSHA regulation for companies using dangerous chemicals. The guidance document of the Exploration and Production Forum of petrochemical companies is included as a second example of this tradition.

14. *US defence/aerospace industry/EG&G.* The MORT system represents a compilation of experience from the defence and aerospace industries into a model of accidents and management systems. It was later developed in Scandinavia to a short version (SMORT).

15. *Scandinavian school of participative management.* An extensive tradition of using sociotechnical principles in design of work and the involvement of the work group in safety control. A number of papers at a recent Scandinavian conference, which brought together researchers from a range of research groups, is used to illustrate this approach.

16. *Sociological research.* The tightly specified sociological theory of accident causation produced by Dwyer is chosen as representative of this research approach. It links accidents as socially produced errors to the organization and reward systems of companies.

17. *Confederation of British Industry.* A survey among member companies into the factors relevant to successful SHE management.

18. *INSAG.* Principles and audit mechanisms for 'safety culture' developed by the International Atomic Energy Authority after Chernobyl.

19. *Berlin.* Analysis of safety organization in German nuclear plants based on incident analysis.

20. *TNO.* Development of an analysis and audit system as a basis for certification of company management systems.

21. *Brookhaven/Pennsylvania State*.   NRC-funded work to develop high-level safety indicators. Development of a six-scale audit with 20 factors based on literature. Validation studies.

22. *Leiden/Manchester*.   Development of a safety management system (TRIPOD) for Shell, growing out of work on cognitive failure to develop a classification of eleven general failure types linked to organizational factors.

23. *Delft*.   Modelling of SHE management as a problem-solving activity, based on studies of safety services, management in the steel industry and the regulatory role of government.

24. *Safety climate scales*.   A range of studies, often inspired by the pioneering work of Zohar, to develop scales of attitudes, perceptions and actions related to safety performance.

25. *Surrey*.   Studies for insurance companies, the chemical and steel industries in the UK to develop safety climate scales based on facet theory.

26. *Kent*.   Studies of environmental climate in companies in the paper industry.

27. *VTT/Finnish nuclear industry*.   Studies in the Finnish nuclear industry on the role of operators in forming safety culture.

28. *Risø/World Bank*.   Theoretical writings, partly inspired by the proceedings of the World Bank symposia, criticizing current approaches to organizational factors.

29. *Berkeley*.   Studies of high reliability organizations, particularly the flight operations on US nuclear aircraft carriers, to build new theory on on-line risk management of complex, dynamic work processes.

30. *Perrow*.   Normal accidents theory developed by Perrow in 1984 and tested and developed by others, e.g. Meshtaki. A test of the predictive value of this theory against HRO theory was carried out recently by Sagan.

31. *(North) London*.   Studies deriving from the sociotechnical tradition and from Turner's analyses of major disasters.

32. *NIA*.   Dutch Institute of Working Conditions studies of the integration of management systems for safety and health, environment and quality, based on organizational learning theory.

33. *Management standards*.   Development of national and international standards for safety and health, and environmental management based on ISO quality management standards.

34. *Trondheim*.   In-depth studies of organizational dynamics and of the introduction of internal control systems into the Norwegian on-shore industry, taking a developmental perspective.

11.4.2 Categorizing the studies

Table 11.1 provides a classification of the major studies, traditions and theoretical approaches. The principal industry or activity within which, or for which, the work was conducted is given, together with the country of origin of the researchers and, in brackets, the country in which the work was carried out, if that is different. The coding of the purpose of each approach is as given earlier (11.3). The main product or intended product of the research is also indicated in the table.

The following classifications consider the research data and methods used. The data sources are classified as:

1. Accident analysis:
   (a) case studies
   (b) statistical analysis of large samples
2. Field study/questionnaires of company data:
   (a) case studies
   (b) statistical analysis of (matched) samples
3. Questionnaire/interview data of employees/managers attitudes/beliefs/perceptions
4. Expert opinion
5. Theory/literature
6. Simulator studies.

Where the studies were experimental the analysis method is mentioned. Most of the terms are self-explanatory. The term 'case study' is used to indicate a study where data were collected from one or a small number of companies and subjected to analysis to reveal patterns of factors on a qualitative, rather than quantitative, basis.

If an explicit theory or body of methods was used this is indicated. These may be theories outside SHE management or explicit models of the SHE management system. A blank means that the approach was essentially pragmatic, with no explicit links to a framing theory.

'Topics' indicates whether the research has generated a list of relevant organizational factors that the authors appear to claim as 'complete', 'partial' or as 'suggestions', or, in the case of experimental studies, as having a 'proven' relevance or irrelevance to successful SHE management. The classification is explained further in the next section. It classifies the topics found or suggested by the research into the four frames of Bolman and Deal (1984):

St = structural
H = human resources
P = political
S = symbolic

Where only one or two topics fall into a given frame, it is placed in brackets. Where the letter is missing, there are no topics in a given frame.

The entry under 'Stage of work' gives an idea of how extensive and far developed the line of study is. Dates give the approximate spread of publications. If only a single date is given, the work is still current; an end date indicates that the group/line of research is no longer active, either because it has clearly been rounded off or because there have been no publications located for a period of at least four years.

The main publications under each approach were read and lists made of the topics which they have showed or claimed to be relevant for successful SHE management. In many cases such a list could be easily made since the authors themselves arrived at an explicit list in one or more of their papers. In other cases a certain amount of interpretation was necessary to characterize the main themes. Some studies or publications, such as commercial audits and checklists, contain many hundreds of specific questions. We have not made our classification of those types of

Table 11.1 Type and approach of organizational studies

Study number	Group/type	Purpose	Industry	Country	Data sources	Research design	Explicit theory	Topics	Stage of work
1	UCLA	1 Assessment tool	Nuclear	US	5	—	PRA, task performance, two level	Partial (from 6) St (H) P	Demonstrative case application 1990
2	Birmingham	1 Assessment tool	Nuclear/ process	UK	4, 5	—	Human reliability analysis	Suggestions (St) H, P (S)	Speculative 1994
3	EDF	1, 4 Assessment tool, climate scale	Nuclear	France	3, 5, 6	Case study Acc. modelling	Organizational culture, clinical stress	Complete St, H, P, S	Field application 1988
4	Technica, Four Elements	1 (3) Assessment tool, audit	Chemical	UK/NL (Gr, P)	1b, 2a	Case study Acc. analysis	Feedback loops, systems	Complete St (P, H)	Demonstrative case application 1988
5	Science Applications/ Maryland	1, 2 Assessment tool	Nuclear	US	5	—	Lewin, PRA, task performance, six levels	Complete St, H, P (S)	Demonstrative application 1988
6	Minnesota	2, 5 Assessment tool, theory	Nuclear	US	2b	Correlational	Business economics, three levels	Complete St, H, P (S)	Completed analyses 1987–90
7	NIOSH *et al.*	3 (4) Checklist, theory	Broad spectrum, mines, manufacturing	US, Canada, UK, S.E. Asia	2b	Matched high/ low accident group	—	Proven St, H, P, S	Multiple studies 1977
8	Imperial College	3 Model	Chemical, retail, construction	UK	2a, 5	Case study	Deming circle (conflict)	Suggestions St, H (P, S)	Completed analyses 1982–88

No.	Source	Type	Industry	Region	Stage	Method	Theory	Completeness	Status
9	Commercial auditors	3 Audit	Various (large company)	World-wide	4	Case study (correlational)	Deming, systems	Complete St, H (S)	Commercial application 1960s
10	Health and Safety Executive	3 Model	Various	UK	4, 1b	Limited epidemiology	Deming	Complete St	Advisory documents 1981–92
11	(Dutch) Government	3 Legal frame Audit	Various	NL	4, 5	—	—	Complete St, H (P)	Guidance note 1994
12	US management texts	3 Model	Various	US	1b, 4, 2a, 3	Case study	Accident sequence	Principles St, H, P, S	Textbooks 1931
13	Chemical process industry	3 Audit tool	Chemical	US	4, 5	—	Systems, Deming	Complete St, H, P, S	Approved guidelines 1993
14	US defence and aerospace	3 Assessment tool	Mainly high hazard	US	1a, 4, 5	Case study	Haddon	Complete St, H, P (S)	Validated tool 1980
15	Scandinavian, participative management	3, 4 Theory, tools	Various	Scandinavia, US	2a, b, 3, 5	Correlational, case study	Sociotechnical, Lewin	Partial St, H, P (S)	Experimental validation 1980s
16	Sociological (Dwyer)	3 Theory	All	Brazil, Fr, NZ	1a, 2a, b	Case study, regression	Sociological	Complete St, H, P, S	Completed book 1978
17	Confederation of British Industry	3, 4 Checklist	Various	UK	2b	Case study	—	Complete St, H (P) S	Completed survey 1989–90
18	INSAG, OSART/ASCOT	3, 4 Audit	Nuclear	World-wide	4, 5	—	—	Complete St, H, P, S	Operational audit 1986

Table 11.1 (cont)

Study number	Group/type	Purpose	Industry	Country	Data sources	Research design	Explicit theory	Topics	Stage of work
19	Berlin	3, 4 ?Model, audit	Nuclear	Germany	1a, b	Case study	—	Complete St, P (H, S)	Data collection 1991
20	TNO, SMART	3, 4 Audit	Chemical	NL	2a, 4, 5	Case study	Mintzberg, Deming, Harrison	Complete St (H, P) S	Demonstrative case application 1990
21	Brookhaven, Pennsylvania State	3, 4 Audit and assessment tools	Nuclear	US	5, 2a	Case study	Mintzberg	Complete St (H, P) S	Completed 1988–94
22	Leiden/ Manchester	3, 4 Audit, incident analysis tool	Chemical	NL, UK (world-wide)	1a, 2a, 4, 5	Case study, factor analysis	Accident sequence	Complete St, H, P, S	Validated field application 1989
23	Delft	3 (4) Model, training	Chemical, steel	NL	2a, b, 3, 5	Case study before and after	Systems, SADT, deviation, life cycle, three levels	Complete St (H, S)	Demonstrative research application 1985
24	Safety climate, scales	4 Climate scales	Various	Israel, US, F, UK, NL, Sp, EU	3	Factor analysis	—	Complete H (P) S	Multiple studies and scales 1980
25	Surrey/ Liverpool	4 Climate scale	Steel, chemical	UK (+NL, Lux, P)	3, 2a	Least space analysis	Facet theory	Complete (St) H, S	Experimental questionnaire 1984
26	Kent	4 Climate scale	Paper	UK	3	Principal component, correlational	Organization culture (O'Riordan)	Partial St, H (P, S)	Experimental 1993

No.	Origin	Code/type	Sector	Country	Code	Method	Theoretical basis	Scope	Status
27	VTT/Finnish nuclear industry	4, 5 Model	Nuclear	Finland	3, 2a	—	Perrow	Partial St (H, P, S)	On-going 1991
28	Risø, World Bank	4, 3, 5 Model, theory	High technology	Denmark (world-wide)	5	—	Systems, accident migration	Complete St, H, P (S)	On-going theoretical 1980
29	Berkeley	4, 3, 5 Theory	Flight operations, energy grid, ATC	US	2a, 5	Case study	Group dynamics, Mintzberg, Aristotle	Partial St, H, P, S	Completed empirical study 1987–92
30	Perrow	4, 5 Theory	High technology	US	1a, 2a, 5	Case study	Systems	Partial St, H, P	Book + validation 1984
31	North London, London	4, 6 Theory, model	High technology	UK	1a, 2a, 5	Case study	Perrow, sociotechnical	Suggestions St, H, P (S)	Speculative 1978
32	NIA	5 Theory, model	Various	NL	2a, 5	Case study	Organizational learning	Partial St, H (P) S	Completed empirical study 1994
33	Standards, ISO/national	5, 3 Checklist, audit	All	World-wide	4	—	Quality management	Complete St, H (P, S)	Ongoing standardization 1991
34	Trondheim	5, 6 Theory, model	Various	Norway	2a, b	Case study, correlational	Bolman and Deal	Partial St (P)	Completed empirical study 1990

DK = Denmark; EU = European Union; F = Finland; Fr = France; Gr = Greece; Lux = Luxembourg; NL = The Netherlands; NZ = New Zealand; P = Portugal; Sp = Spain; UK = United Kingdom; US = United States; Acc = accident.

study at the question level but at the level of the scales or subdivisions of the checklist, to indicate the emphasis they place. The same applies to the safety climate studies, which use questions to form scales and then perform factor analysis to discover whether their scales are robust; we have used the level of proven factors, not initial scales. Finally studies are often partial; because a topic is not mentioned by an author, it does not mean that the topic is not relevant nor even that the author would not consider it relevant.

The organizational factors mentioned by the different lines of research were tabulated and classified under headings. This process is inevitably a somewhat subjective one. Different authors use different terms to mean roughly the same thing, or at least what these reviewers have interpreted as the same thing (see also Shannon *et al.*, in press), so that allocation to a broad category contains some interpretation from ourselves. The total number of times a topic is mentioned in the subsequent tables should not therefore be interpreted too rapidly to indicate its degree of importance, though we shall interpret it as some indication of degree of agreement. Low numbers of mentions can arise from a lack of attention up to now in research.

11.4.3 Classification of organizational factors: Bolman and Deal

In order to give some ordering to the topics, the four frames of Bolman and Deal (1984) were used. These frames represent four different perspectives of an organization, which accentuate four different ways of looking at it and at what goes on inside it. Bolman and Deal derive these perspectives from the organizational literature to typify four different schools of research and thought, which have concentrated on rather different aspects of the organization in their studies. They do not claim that any one perspective is more correct than another, simply that a given perspective may provide a more useful view of certain sorts of problem in certain sorts of situation or organization (Bolman and Deal, 1984, p. 250). This is rather like staining cross-sections of a specimen on a microscope slide with four different stains to accentuate different structures within it. The four frames are characterized in Table 11.2 with typical keywords to indicate the focus of the perspective. The purpose of using this classification is to throw some light on, and generate debate about, the dominant perspectives in SHE management. Such debate has been raised by Rasmussen (Rasmussen and Batstone, 1990) and by Rochlin (1989) among others. It also lies at the heart of discussion about the appropriateness of bureaucratic, rule-based SHE audit systems (usually constructed from a largely structural perspective) for some (if not all) types of organization.

Since the four frames do not subdivide the organization and its functioning into exclusive parts, but accentuate different aspects of all the parts, some care must be taken in allocating topics to one or another frame. This is dependent on choosing the correct level of analysis. For example, it is not appropriate to take the activity 'meetings', which can be seen from all perspectives (Bolman and Deal, 1984, p. 247): in the structural frame as a place to formalize decisions, in the human resources frame as a place to share feelings, in the political frame as a place to score points and bargain, and in the symbolic frame as a place to enact rituals and to reinforce or change culture. The level of analysis must be in terms of what goes on, or what the researchers/authors suggest is intended to go on at the meeting. In analyzing the

Table 11.2 The four frames of Bolman and Deal (1984)

Frame	Assumptions	Keywords
Structural	Organization exists to accomplish goals Structure should fit goals, environment and technology Rationality should be maximized Specialization and structure are basic principles Coordination through hierarchy and rules	Responsibility, hierarchy, structure, rules, feedback, command and control, deviation control, organizational goals
Human resource	Organizations exist to serve human needs There is a mutual dependency between organizations and people Prime goal is organizational fit	Competence, group dynamics, participation, needs and motivation, achievement, learning, leadership style
Political	Decisions in organizations are about allocating scarce resources Organizations are coalitions between groups with different values, beliefs and realities Goals evolve from bargaining and negotiation Power and conflict are central issues	Power, influence, conflict, bargaining, game theory, negotiation
Symbolic	Many significant events in organizations are ambiguous and not subject to one rationality Interpretation and meaning are crucial issues Symbols reduce uncertainty and release creativity Strong cultures get results	Values, symbols, myths, stories, heroes, role playing, scenarios, culture

literature there is an inescapable element of subjectivity in interpreting what authors intend for the purposes of such a classification. Nevertheless the interpretation is in most cases reasonably clear from the authors' own words, or from the context, as to where the emphasis lies. It is also open to authors and other researchers to dispute the classifications made in the tables.

11.5 SOME QUANTITATIVE RESULTS

The data from Table 11.1 is summarized in the following sections to illustrate some quantitative findings about research in this area, particularly in respect of the national and industry base it has, and its lack of empirical basis. The relative youth of the field can be seen in the dates at which different lines of research were started. Only five date from before 1980; nine were started between 1980 and 1985, 12 more between 1986 and 1990, and the remaining eight began after 1990.

11.5.1 Industry and national provenance of research

Table 11.3 gives the cross-tabulation of the industry where the work was done and the purpose of the study. Totals are smaller than the sum of the rows or columns because many studies covered several purposes or industries. For the full definition of each purpose see above. The expected dominance of high hazard industry in general, and particularly in the attempts to quantify management factors and discover high level indicators for regulatory intervention, is confirmed.

Table 11.4 gives the country of origin of the studies. It shows the dominance of North America and northwest Europe, though this may partly be biased by the languages used in the search for papers. However, a search of conferences in this area seems to indicate that there is more to it than that. Management as an academic subject seems to be an invention of that area of the world, and hence safety management also.

11.5.2 Research method and data sources

Most research lines (25) have conducted at least some field work. Half (17) have used secondary sources and half (17) have aimed at development of methods or models. The data sources used are shown in Table 11.5.

Nine research lines are based on little or no field work, eleven almost entirely on case studies and only fourteen are empirical in some way. Of those we pay particular attention in the further discussion to the results of the studies that have so far

Table 11.3 Purpose by industry

Industry	Management in PRA	High level indicator	System structure	Climate, culture	Development integration	Match to original type	Total
Nuclear	4	2	3	5	2	—	9
Chemical, high hazard	2	—	9	8	3	1	12
General	—	—	11	5	3	1	16
Total	5	2	21	18	8	2	34

Table 11.4 Country of origin

Country	Number
N. America	12
UK	11
The Netherlands	7
Scandinavia	5
France	2
Germany	1
World-wide	3
Other	3

produced empirical evidence to show that organizations with particular characteristics are better at managing SHE. These are limited to EDF, Minnesota, NIOSH *et al.*, participative management, Dwyer *et al.* and Surrey (studies 3, 6, 7, 15, 16, 24, 25), which looked at relationships with accident rates, Kent (26), which has empirical data on the relationship with environmental performance, Four Elements and HSE (4 and 10), which looked at management factors involved in accidents/incidents, and Trondheim (34), which studied empirically management factors in relationship with success in implementing internal control systems. The topics where proof is available from these studies are given in heavy type in Tables 11.6 to 11.9. This can be proof that the topic is positively or negatively related or unrelated to SHE performance in the study. See the discussions for details.

Table 11.5 Data sources

Data source	Number
Accident cases	6
Accident statistics	4
Company cases	16
Company statistical	7
Individual questionnaire	8
Expert opinion	11
Theory, literature	19
Simulator	1

11.5.3 Factors identified by studies

The Tables 11.6 to 11.9 summarize the topics that the different research lines indicate or prove to be important in a successful SHE management, classified according to the four frames set out in Table 11.2. The terms used are generally self-explanatory. The brief descriptions under each frame provide more information where this is not the case.

Structural factors

Table 11.6 shows the structural topics. Measurable goals, competence in organizational development and access to external expertise and available resources are all associated with success in developing internal control systems in the Trondheim studies (34). Review activities and learning systems in design, maintenance and construction and operations, and task checking and inspection are factors found among significant numbers of accidents in study 4.

Topics positively related to low accident rate or good environmental performance in at least one research study under 6, 7 or 26 are the availability of (financial) resources, a problem-solving approach, a stable workforce, safety training systems, good communication channels, good coordination and centralization of safety control, a specialist safety service in the company, good records, a small span of control for, and time to plan by, supervisors, and risk assessment as an activity. The

Table 11.6 Structural topics

Topic	Mentioned in studies	No. of studies
1. **Goals, standards**, performance indicators	1, 4, 5, 6, 9, 10, 11, 12, 13, 17, 22, 23, 27, 28, 29, 30, 32, 33, **34**	20
2. Policy, plans	4, 5, 9, 10, 11, 12, 13, 14, 16, 17, 18, 22, 23, 26, 29, 30, 32, 33	18
3. Responsibility, authority, structure, hierarchy	1, 2, 3, 4, 5, 6, 8, 9, 10, 12, 13, 14, 15, 16, 17, 18, 20, 21, 22, 26, 29, 30, 33, 34	24
4. **Resources (general)**, proactive	1, 4, 6, **7**, 8, 11, 13, 14, 17, 18, 20, 22, 23, **26**, 30, 33, **34**	18
5. **Human resources planning and systems:** selection, **turnover, training**	1, 4, 5, 6, **7**, 9, 11, 12, 13, 14, 15, 16, 17, 18, 19, 20, 21, 22, 23, 28, 29, 30, 31, 32, 33	25
6. Information flow, **communication channels**	1, 3, 4, 5, 6, **7**, 8, 9, 11, 13, 14, 15, 16, 17, 18, 19, 20, 21, 22, 23, 27, 28, 29, 30, 31, 33	26
7. **Rules, procedures, formalization**	1, 4, 5, **6**, **7**, 9, 11, 13, 14, *16*, 17, 21, 22, 27, 28, 29, 30, 31, 32, 33	20
8. **Coordination, centralization**	1, 5, **6**, 13, 16, 21, 22, 26, 27, 28, 29, 30, 33	13
9. **Specialist advisory service**	5, **7**, 8, 11, 18, 31, 32, 33, 34	9
10. **Supervision, task check/monitor**	**4, 7**, 9, 11, 12, 13, 14, 15, 18, 20, 32	11
11. **Review, evaluate, feedback, audit**	1, 3, **4, 6**, 8, 9, 10, 11, 13, 14, 15, 17, 18, 19, 20, 21, 23, 25, 28, 29, 32, 33, 34	23
12. Organizational learning, change	**1, 4**, 6, 9, 11, 12, 13, 14, 15, 17, 18, 19, 20, 22, 26, 28, 29, 30, 32, **34**	20
13. **Risk assessment/recognition, incident/ accident analysis, inspection**	2, 4, **7**, 9, 10, 11, 12, 13, 14, 17, 19, 20, 21, 22, 31, 32, 33	17
14. Deviation vs. *boundary* control	4, 8, 9, 10, 12, 18, 19, 23, 28, 30, 31, 34	12
15. **Problem solving**/Deming circle	1, 6, 8, 9, 10, 12, 13, 20, 23, **26**, 32, 33	12
16. Specific link to **life cycle** phases (**design review, maintenance**, contractors, **operations**, materials control, emergency preparedness)	1, **4**, 9, 10, 13, 14, 17, 18, 20, 22, 23, 25, 28	13
Total		281
Average per topic		17.7

presence of accident reporting and analysis systems was found in one study under 7 to be unrelated to accident rate, but evaluation and review systems were found in other studies to be associated with success.

On the subject of rules the studies are most contradictory; there are studies under 6, 7 and 26 which find a positive correlation between the quality and quantity of rules and good performance, but others under 7 find no correlation. The studies in italics (17, 27, 28, 29) warn explicitly against imposing centrally made rules on work groups. The difference of opinion would appear to relate to the type of risk management (complex on-line management requires groups to evolve and con-stantly revise and reconfirm their own rules, in contrast to off-line management) and possibly to the type of organization studied or the state of development of its system of rules.

Many studies advocate systems for deviation control, modelled on quality man-agement systems. Research lines 4 and 28 explicitly reject this and argue that it is not strict adherence to a norm that should be the aim, but guarding the management system from approaching dangerous boundaries, while leaving the employees free-dom to choose within those boundaries.

Human resource factors

Table 11.7 presents the human resources topics. In the area of communication these emphasize the content and quality of the communication as opposed to the presence of communication channels (structural). In the area of training they emphasize the result of the training (competence) rather than the presence of training programmes (structural).

All of the topics in bold have been found to be related to low accident rate and/ or good environmental performance in at least one of the studies reviewed, apart from involvement of family, which was found to be unrelated in all studies in which it was looked at. The specific leadership style found to be related to low accident rate was one linked to interpersonal skills and perceived concern for the group and its dynamics.

Political factors

Table 11.8 contains the topics from the political frame of Bolman and Deal. Com-pany profitability and availability of resources were found to be positively related to high safety in two studies but unrelated in a third. The absence of incentive payment schemes for production and the presence of safety incentive schemes were found to be related once each to low accidents. Sanctioning of violations was related to high accidents. One study found a correlation between use of discipline and low acci-dents, while others found that low accident companies were characterized by the use of counselling rather than discipline (which was associated with high accident com-panies), while a further study found use of discipline to be unrelated to safety performance. Openness to criticism, good labour relations, low stress and low grievance rates were all related, as would be expected, to low accident rate, whilst good labour relations were also related to success in introducing an internal control system. External pressure from regulators and the presence of an 'order-seeking' management (as opposed to a crisis management approach) were related to good performance.

Table 11.7 Human resources topics

Topic	Mentioned in studies	No. of studies
1. **Participation, empowerment, innovation encouraged**	3, 6, **7**, 8, 9, 12, 13, **15**, 17, 19, 21, 23, 24, **25**, **26**, 28, 29, 30, 32, 32, 33	23
2. **Feeling of control, efficacy, autonomy,** adaptive capacity	2, **3**, **7**, 15, **16**, 17, 18, 22, 24, **25**, 28, 29, 30, 32	14
3. Flexibility, self-awareness	3, 7, 15, 22, 28, 29, 31	7
4. **Informal organization,** team spirit	3, 5, 6, **7**, 8, **15**, 16, 17, 22, 26, 29, 30, 31, 32	15
5. **Group norms,** dynamics and boundaries, groupthink	2, **7**, 12, 15, **16**, 24, 26, 29, 30	9
6. **Interpersonal/group communication**	1, 3, 4, 5, 6, **7**, 9, 11, 13, 14, 15, 16, 17, 18, 19, 20, 21, 22, 24, 27, 28, 29, 31, 33	24
7. Motivation, (excess) zeal	5, 8, 12, 13, 14, 22	6
8. Experiential knowledge	3, 15, 27, 29	4
9. Competence	1, 5, 9, 11, 13, 14, 15, 16, 18, 21, 22, 27, 28, 29, 33	15
10. **Caring,** concern, consideration	3, **7**, 26, 29, 31	5
11. **Leadership style**	3, 5, 6, **7**, 9, 11, 12, 17, 18, 25, 26, 29	12
12. **Social policy, quality of work-life, career progression**	**3**, 5, 6, **7**	4
13. **Family involvement**	**7**, 9	2
14. **Mature/married workforce**	**7**	1
	Total	141
	Average per topic	10.0

Table 11.8 Political topics

Topic	Mentioned in studies	No. of studies
1. **External pressure**, surveillance	4, 6, 8, 12, 14, **16**, 18, 19, 20, 28, 310, 32	13
2. Company violations, compliance	5, 6, 28, 30	4
3. Urgency, priority of SHE	2, 12, 21, 30	4
4. **Profitable, resources made available**	1, 5, **6, 7**, 12, 13, 16, 29, 30	9
5. Conflict of interest, trade-off, long-term strategy, **order seeking *vs.* damage limitation**	1, 2, 4, 7, 8, 14, 15, 16, 17, 19, 20, 22, 24, **26**, 27, 28, 29, 30, 33	19
6. Slack in system, **workload**, overtime	1, **3**, 6, 12, 16, 18, 22, 28, 29	9
7. **Reward, incentive system**	1, 5, **7**, 12, 13, 14, 15, **16**, 17, 18, 31	11
8. **Sanctioning rule violations**	2, **7**, 16, 22, 28, 29, 30, 31	8
9. **Counselling *vs.* discipline for violation**	**7**, 16	2
10. Freedom of speech, **openness to criticism**	3, **7**, 13, 18, 21, 24, 28, 29, 30, 31	10
11. **Trust, good relations**, fairness	2, 3, 5, **7**, 15, 29, 30, 31, 32, **34**	10
12. **Presence/activity of safety committee/union**	**7**, 11, 16, 33	4
	Total	104
	Average per topic	8.7

The evidence on the relationship of unions and safety committees to safety performance is much more equivocal. On balance there is a positive relationship between having a trained and balanced membership of committees and safety (including the presence of the safety officer on it) as opposed to the opposite, but there is no clear relationship between safety and the presence (as opposed to absence) of a committee and one study found a negative correlation between safety and the presence of a union (perhaps reflecting poor labour relations). The evidence on this topic is probably strongly influenced by the country, labour relations tradition and company history.

Symbolic factors

Table 11.9 presents the symbolic topics according to Bolman and Deal's classification. Top management commitment and real visibility in that commitment are found several times positively related to safety, as are supervisor's and individual commitment, importance of safety as a value, the safety attitudes of co-workers and work as a source of pride in the company. High satisfaction with the safety programme is related to high safety, but a high rating of personal safety consciousness was found in one study in low safety gangs. Low accident rate was found associated with attribution of causes to, and attention for, human factors as opposed to an attitude that safety was a purely technical issue. Such a distinction should not be confused with the belief that accidents are due to accident proneness, which is often seen as unalterable and hence unmanageable.

Table 11.9 Symbolic topics

Topic	Mentioned in studies	No. of studies
1. **Top management commitment (time and resource allocation)**, policy as ritual	3, **6**, **7**, 8, 9, 12, 13, 14, 17, 18, 20, 21, 22, 24, **25**, 28, 29, 32, 33	19
2. **Supervisor commitment**	**7**, **24**, 25, 29	4
3. **High status safety officer**	**7**, 12, 24	3
4. **Standards of excellence, self-satisfaction**, fix-it/crisis mentality	2, **3**, **7**, 13, 17, 22, **24**, 33	8
5. **Importance of safety/environment**, reality of risk, macho culture	2, **7**, 13, 16, 17, 18, 20, 21, 22, 24, **26**, 29, 31, 33	14
6. Common goals, **attitudes of colleagues**, ownership	2, 15, 16, 17, 18, 21, **24**, 25, 29	9
7. **Personal commitment**	3, **7**, 12, 13, 16, 17, 24, 27, 32	9
8. **Work as a source of pride**	**3**, **7**, 16, 22, 29, 32	6
9. Morals, fairness	3	1
10. Rumours, stories about safety	29	1
11. **Attribution to human factor**, multicausality	**7**, 12	2
12. Safety culture, fit of system to organizational culture	5, 6, 12, 19, 20, 21, 22, 23, 27, 28, 32	11
13. **Safety promotion**	**7**, 9, 12, 13, 17	5
	Total	92
	Average per topic	7.1

Table 11.10 Attention to topics by research lines

Research lines	No topics in a given frame	One or two topics in a given frame	Three or more topics in a given frame	Average number of research lines per topic (Tables 11.6–11.9)
Frame				
Structural	1	2	31	17.7
Human resources	2	7	25	10.0
Political	4	13	17	8.7
Symbolic	6	14	14	7.1

The evidence on safety promotion is mixed: one study found attention to promotional campaigns to the exclusion of other activities correlated with high accidents; another study found a positive correlation between the presence of promotional campaigns and safety. This is likely to be an effect of the stage of development of the SHE management programme in the two studies.

Aggregated results

When we aggregate the topics mentioned across the lines of research that we surveyed, we arrive at Tables 11.10 and 11.11. Table 11.10 shows how many research lines have studied topics from each of the frames. The dominance of studies on, and

Table 11.11 Empirical proof of links between topics and company performance

Topics	Safety/ environmental performance	Other empirical indicator	No empirical indicator	Total number of topics
Frame				
Structural	10	7	3	16
Human resources	10	—	4	14
Political	10	1	2	12
Symbolic	10	—	3	13
Total	40	8	12	56

the degree of agreement about, topics in the structural frame is clear. The other frames follow in descending order of attention and concensus.

Table 11.11 presents the evidence on the number of topics that has been linked in one way or other to good or poor safety or environmental performance. The evidence here is very different. The proof of links between topics in the four frames is absolutely evenly distributed.

11.6 DISCUSSION AND CONCLUSIONS

At the global level of analysis presented in this paper a number of points emerge.

11.6.1 Atheoretical nature of research

Many lines of research appear to be either entirely atheoretical or to be based on little more than a methodology linked to an explicit model of the accident process, to general systems theory or to the 'plan–do–check–modify' cycle derived from quality management (Deming, 1990). These are little more than principles for ordering and systematizing SHE management systems, in particular their structural elements. Only sociotechnical theory, the theory of Lewin, the classification of organization types from Mintzberg (1980) and the ideas of organizational culture, notably those of Hofstede (1986), and more recently the theories of the learning organization (e.g. Senge, 1990) seem to have penetrated the field from organization theory, but none of them have been extensively developed or tested.

The field itself has generated some theory, notably that of Perrow (30) and of the High Reliability Organization (29), which have been explicitly tested against each other by Sagan (1993), in respect of US nuclear defense activity, a study coming down in large part in favour of Perrow. For the rest the study of SHE management seems to have been conducted in something of a theoretical limbo.

11.6.2 Empirical development

The field is still young and the number of empirical lines of research and studies carried out is still very small. The emergence of management as a topic of SHE research cannot really be traced to any sudden increase in management failure. It is likely that it has much more to do with a process of maturation in the field, coupled with the changes in emphasis of regulatory activity. As the technical aspects of SHE that had dominated the first age of the subject became more known, questions were raised about why these were not implemented by the human operators in the system. That led to the second age, concentrating on understanding individual failure and the adaptation of technology to the strengths and limitation of operators. As knowledge in this area advanced, the questions moved to asking why companies were more or less successful in implementing this new knowledge, and the third age was born.

The limited empirical evidence available that links management and organizational characteristics to safety or environmental performance indicates how far we have to go in establishing a firm basis for the enormous amount of work and activity that is now going on world-wide. Most of the links that have some empirical support are based only on one or two studies and the evidence on others is at best equivocal. A number of topics, such as policies and plans, responsibility and structure, competence and motivation, which are cited as essential by large numbers of authors, have no empirical basis to back this importance. A vast amount of the state of the art rests on an insubstantial tissue of expert opinion and the analysis of case studies, whose generalizability has not yet been proven.

Cohen (1977), in one of the earliest studies in this field, classified the evidence available to him into three sorts: what SHE experts, managers or workers perceive as important for achieving SHE, what good companies do to manage safety and what matched pairs comparisons of good and bad companies show to be significant differences between them. He warned of the care needed to draw conclusions from studies of these different designs. All that good companies do in the name of SHE is not necessarily crucial to their performance. What distinguishes them from poor companies may only be a fraction of that total. That fraction may also change. What distinguishes a terrible from a mediocre company may be very different from what distinguishes a mediocre from a good, or a good from a superlative company. There is at least some evidence from studies under heading 7 that Bolman and Deal's structural topics are critical at the lower end of the performance scale but that they cease to discriminate once a company is at least mediocre.

What experts and managers perceive to be important is likely to show even less overlap with the crucial factors that discriminate good from poor performance. The tables show that there is the most concensus between studies over the structural topics and yet a higher percentage of them than of the topics from the other frames are not (yet) correlated with safety performance. It is true that some of those topics are linked to success in implementing a SHE management system but the overall standard of proof is thin. It is therefore of particular concern that most of the audit systems that are routinely used in all branches of industry rest on this basis of conventional wisdom. Only the studies of Four Elements (4) can claim some empirical support but this is limited to support from accident analysis, not from correlational studies with safety performance. This is not to deny the enormous problems in carrying out validational studies making comparisons between companies. They must have sufficient accidents to be able to show significant differences and differences in reporting rate must be distinguished from differences in accident rate. However, it is of concern that, since the pioneering studies of NIOSH, very few such studies seem to have reached the literature.

The other main type of empirical study is that developing so-called 'safety climate' scales (24, 25, 26). Only two research lines have made any attempt at external validation of their scales (25, 26) and the published evidence on both is, as yet, minimal. The rest (grouped under heading 24) still show comparatively little consistency between themselves and their relationship with the proven results from the other empirical studies is far from clear.

Having made this plea for empirical research as the ultimate way of proving the value of specific management structure and practices, we must also add a note of caution, and even of theoretical doubt. The status of management as a science is not clear. Safety problems are in many ways social constructs, depending for their definition and solution just as much on how people perceive them as on any objective characteristics. There are many ways of managing a given technology, as the literature on management culture demonstrates. It is therefore not to be expected that clear-cut links can or will be found between individual management characteristics and safety performance. Other criteria for judging the appropriateness of management, such as acceptance by the managed, fit to company culture, or the feeling of insight a theory gives to managers in applying it to their own company, may be equally important. This leads us on to our next point.

11.6.3 Definitions of safety culture and climate

The terms 'safety (or SHE) culture' and 'organizational culture' have been used extensively in this chapter. Unfortunately the existing literature is irrevocably confused in its use of the term 'safety culture' (see also Williams, 1991). There is some agreement that the term 'culture' refers to the shared beliefs, attitudes and norms that govern how the company is run, how it sets its priorities, how it communicates and cooperates internally and externally. Companies within the same industry differ in these, particularly across countries (e.g. Hofstede, 1986). If we apply this definition to safety, then 'safety culture' should be the shared beliefs, etc. relevant to safety and its management. This is the way that ACSNI (1993) used the term safety culture, applying it to all beliefs about SHE, including giving it a low priority. Some writers only talk of a 'safety culture' if the achievement of safety is a *high* priority in the shared beliefs. This seems to be an incorrect use of the term. We have tried to use the term 'safety (or SHE) climate' to refer to scales to measure this dimension, though it is clear that some of the scales also incorporate measurements that are not drawn from the attitude or belief sphere. Examination of tables 11.6 to 11.9 shows that the studies we have labelled as 'safety climate' research tend to tap into the symbolic frame of Bolman and Deal to a greater extent that many other research lines, particularly those concerned with system audits, which concentrate on the structural frame. It is tempting to propose that safety culture should therefore be used to refer to the three non-structural frames, though the overlaps between frames are still quite large.

We would therefore plead for a definition that would reserve the term 'safety culture' for the different ways in which companies in the same industry or activity fill in the structural frame to arrive at different ways of managing safety.

11.6.4 Lack of links to organizational types and to the dynamics of management development

Sometimes the term 'safety culture' is mixed in with the notion of organization type (e.g. Mintzberg, 1980; Handy, 1976), a theory which classifies organizations into machine bureaucracy, divisional company, etc. Such a classification is usually linked to the technology and market in which the organization operates. We regard this use of the term 'safety culture' as irrevocably confusing. Organization types are not just different cultures, they also have different structures. It is true, however, that remarkably little attention has been paid to matching SHE management systems to different types of organization, for example classified according to the typology of Mintzberg (1980). The dominance of studies in the nuclear, followed by the chemical, industry is evident. Many of the other studies have been conducted in large organizations. This means that the literature is dominated by studies carried out in high technology, largely bureaucratic organizations, often with very considerable investment in safety and environment (and also in health) and sophisticated defence-in-depth systems. This explains to some extent the great emphasis (and consensus) in the field on the topics in the structural frame. It has led to the idea that SHE management systems must be rule-bound and rule-dominated. Audit systems, certification regimes and regulatory checks reinforce this view with their search for indicators that can be easily detected and proven. Yet the accumulated evidence shows that this limited structural approach misses three-quarters of the factors that have proven links to performance.

Even in the world of highly complex bureaucracies, the studies of the Berkeley school (29) and Perrow (30) have suggested that rules imposed by the organization can be counter-productive and that delegation of control to working groups or on-line managers, intensively communicating with each other, with experiential knowledge and shared, internalized rules, linked to a good understanding of the acceptable boundaries within which the system can function safely, can be a way of achieving greater reliability (Rasmussen, 1993). If we move to organizational types that do not rely for their normal management control on extensive explicit rules, the mismatch with the dominant SHE management paradigm is even more clear. Research organizations, professional bureaucracies, information technology companies, project organizations, many consultants, hospitals and universities, to name a few of the organizations that have often been new entrants to SHE regulation in the last 20 years, do not fit the bureaucratic model but have not been the subject of sufficient study to indicate what SHE management models would fit them. The small companies in traditionally regulated industries also do not have the luxury of being able to afford the paperwork systems implicit in rule-based systems. It is likely that they can rely much more on the effective functioning of topics such as competence and communication, which fall under the non-structural frames. However existing assessment techniques are not very well equipped to measure these.

Finally, the survey reveals the lack of research lines that have concentrated on the dimension of organizational learning and change. One aspect of this is the way in which the company learns to improve its SHE management system. The importance of such a system is generally accepted, as can be read from Table 11.6. There are beginning to be a few studies that use organizational learning theory to look at the introduction, development and integration of SHE management systems. What no studies yet seem to have done is to study the other side of the coin, namely the way in which organizational change undertaken, for example, to retain or improve competitiveness, which is becoming an increasingly important factor in organizational life, impacts on the long-term integrity of SHE management. If organizational change, rather than organizational stability, is to become the central feature in competitive survival, as is increasingly preached, then the bureaucratic model, which learns incrementally from its mistakes in an essentially static environment, is likely to be even less suitable as a model for future companies. Alternatively, if it could be shown that only such stable organizations can be relied on to manage highly dangerous activities successfully, considerations of safety could be used to put a brake on constant change.

In summary, research is needed in all of the following different but overlapping areas representing the articulation of safety climate with organizational type and culture:

1. Is the achievement of high SHE performance in all cultures and organization types dependent on one and the same underlying 'culture of safety (SHE)'?

2. Are there many ways in which SHE can be successfully implemented and managed in any one technology or is there one best way, i.e. can many different successful safety cultures exist?

3. Is the way of achieving a good SHE climate and performance different across different organization types?

References

ACSNI (1993) *Organising for safety*, Health and Safety Commission, ACSNI Human Factors Study Group, third report, London: HMSO.

BIGNELL, V., PETERS, G. and PYM, C. (1977) *Catastrophic failures*, Milton Keynes: Open University Press.

BIRD, F.E. and LOFTHUS, R.G. (1976) *Loss control management*, Loganville, Georgia: Institute Press.

BOLMAN, L.G. and DEAL, T.E. (1984) *Modern approaches to understanding and managing organizations*, San Francisco: Jossey-Bass.

BOOTH, R.T. (1980) Safety: too important a matter to be left to engineers? Inaugural lecture. Aston University. Birmingham.

CARTER, N. and MENCKEL, E. (1990) Group routines for improving accident prevention activities and accident statistics. *International Journal of Industrial Ergonomics*, **5**, 125–132.

CIS (1967) *Human factors and safety*, Geneva: International Labour Office.

DEMING, W.E. (1990) *Out of crisis: quality, productivity and competitive position*, Cambridge: Cambridge University Press.

DEPARTMENT OF ENERGY (1990) *The public enquiry into the Piper Alpha disaster (Cullen report)*, London: HMSO.

DEPARTMENT OF TRANSPORT (1987) *Report of the formal investigation into the sinking of the Herald of Free Enterprise (Sheen report)*, London: HMSO.

(1988) *Investigation into the King's Cross underground fire (Fennel report)*, London: HMSO.

(1989) *Investigation into the Clapham Junction railway accident (Hidden report)*, London: HMSO.

ELDEN, M. (1983) Democratization and participative research in the developing of local theory. *Journal of Occupational Behaviour*, **4**, 21–33.

FORTUNE, J. and PETERS, G. (1995) *Learning from failure: the systems approach*, Chichester: Wiley.

GLENDON, A.I. and McKENNA, E.F. (1995) *Human safety and risk management*, London: Chapman & Hall.

HALE, A.R. (1978) The role of HM Inspectors of Factories with particular reference to their training, PhD thesis, University of Aston in Birmingham.

HALE, A.R. and GLENDON, A.I. (1987) *Individual behaviour in the control of danger*, Amsterdam: Elsevier.

HALE, A.R. and HALE, M. (1972) *A review of the industrial accident research literature*, Committee on Safety and Health at Work, Research Paper 2, London: HMSO.

HOFSTEDE, G. (1986) Culture's consequences: international differences in work-related values. In *Cross-cultural research and methodology series*, eds W.J. Lonner and J.W. Berry, pp. 342–411, London: Sage Publications.

HUISINGH, D. (1989) Cleaner technologies through process modifications, material substitutions and ecologically based ethical values. *UNEP Industry and Environment*, Jan.–Mar. 4–8.

KIRWAN, B. (1994) *A guide to practical human reliability assessment*, London: Taylor & Francis.

KJELLÉN, U. and BANERYD, K. (1983) Changing local health, and safety practices at work within the explosives industry. *Ergonomics*, **26**(9), 863–877.

KJELLÉN, U. and LARSSON, T. (1981) Investigating accidents and reducing risks: a dynamic approach. *Journal of Occupational Accidents*, **3**(2), 129–140.

LEWIN, K. (1951) *Field theory in social science: selected theoretical papers*, New York: Harper & Row.

NORWEGIAN MINISTRY OF LOCAL GOVERNMENT AND LABOUR (1987) *Report: internal control in an integrated strategy for working environment safety*, Report No 1, Oslo: Internal Control Committee.

REASON, J.T. (1990) *Human error*, Cambridge: Cambridge University Press.

ReVELLE, J.B. (1980) *Safety training methods*, New York: Wiley.

ROBENS, LORD, A. (1972) *Safety and health at work: report of the committee 1970–1972*, London: HMSO.

ROBINSON, G.H. (1982) Accidents and socio-technical systems: principles for design. *Accident Analysis and Prevention*, **14**, 121–130.

ROYAL SOCIETY (1992) *Risk: analysis, perception, management*, Report of a Royal Society study group, London: Royal Society.

SAARI, J. and NÄSÄNEN, M. (1989) The effect of positive feedback on industrial housekeeping and accidents: a long-term study at a shipyard. *International Journal of Industrial Ergonomics*, **4**, 201–211.

SENGE, P.M. (1990) *The fifth discipline: the art and practice of the learning organisation*, New York: Doubleday.

SURRY, J. (1969) *Industrial accident research*, Toronto: Ontario Ministry of Labour, Labour Safety Council.

THORSRUD, E. (1981) *Organisation development from a Scandinavian point of view*, Doct. 51/80, Oslo: Work Research Institute.

TRIST, E. and BAMFORTH, K.W. (1951) Some social and psychological consequences of the longwall method of coal-getting. *Human Relations*, **4**, 6–38.

WILLIAMS, J.C. (1991) Safety cultures – their impact on quality, reliability, competitiveness and profitability. In *European reliability*, ed. R.H. Matthews, Amsterdam: Elsevier.

Bibliography

The bibliography contains a number of representative references for each of the 34 lines of research described over and above the specific references cited in the text. Where possible English language papers have been chosen since this publication is in that language. These are arranged by research line in the same order and with the same number as in the text.

1. UCLA

DAVOUDIAN, K., WU, J.-S. and APOSTOLAKIS, G. (1994) Incorporating organisational factors into risk assessment through the analysis of work processes. *Reliability Engineering and System Safety*, **45**(1–2), 85–105.

(1994) The work process analysis model (WPAM). *Reliability Engineering and System Safety*, **45**(1–2), 107–125.

2. Birmingham

KENNEDY, R. and KIRWAN, B. (1995) The failure mechanisms of safety culture. International Topical Meeting on Safety Culture in Nuclear Installations, IAEA, Vienna, pp. 281–290.

(1996) The safety culture HAZOP: an inductive and group-based approach to identifying safety culture vulnerabilities. In *Probabalistic safety assessment and management*, eds P.C. Cacciabue and I.A. Papazoglou, pp. 910–915, Berlin: Springer-Verlag.

3. EDF

LLORY, M. and LARCHIER-BOULANGER, J. (1988) A turning point in human factors studies. *IEEE Fourth Conference on Human Factors and Power Plants*, Monterey, California.

MONTMAYEUL, V., MOSNERON-DUPIN, F. and LLORY, M. (1994) The managerial dilemma between the prescribed task and the real activity of operators: some trends for research on human factors. *Reliability Engineering and System Safety*, **45**(1–2), 67–73.

4. Technica/Four Elements

BELLAMY, J.L., WRIGHT, S.M. and HURST, W. (1993) History and development of a safety management system audit for incorporation into quantitative risk assessment. *International Process Safety Management Conference and Workshop, San Francisco, California, Part II*; September.

HURST, N.W., BELLAMY, L.J., GEYER, T.A.W. and ASTLEY, J.A. (1991) A classification scheme for pipework failures to include human and sociotechnical errors and their contribution to pipework failure frequencies. *Journal of Hazardous Materials*, **26**, 159–186.

PITBALDO, R.M., WILLIAMS, J. and SLATER, D.H. (1990) Quantitative assessment of process safety programs. *Plant/Operations Progress*, **9**(3), 169–175.

5. Science Applications/Maryland

ANDERSON, N.S., SCHUURMAN, D.L. and WREATHALL, J. (1990) A structure of influences of management and organisational factors on unsafe acts at the job performer level. *Proceedings of the Human Factors Society*, 881–884.

MODARRES, M., MOSLEH, A. and WREATHALL, J. (1994) A framework for assessing influence of organization on plant safety. *Reliability Engineering and System Safety*, **45**, 157–171.

SCHUURMAN, D.L. and KRAMER, J.J. (1990) Overview of management and organisational effects on industrial safety. *Proceedings of the Human Factors Society*, 868–870.

6. Minnesota

MARCUS, A.A. (1988) Implementing externally induced innovations: a comparison of rule-bound and autonomous approaches. *Academy of Management Journal*, **31**(2), 235–256.

NICHOLS, M.L. and MARCUS, A.A. (1990) Empirical studies of candidate leading indicators of safety in nuclear power plants: an expanded view of human factors research. *Proceedings of the Human Factors Society*, 876–880.

OSBORN, R.N. and JACKSON, D.H. (1988) Leaders, riverboat gamblers, or purposeful unintended consequences in the management of complex dangerous technologies. *Academy of Management Journal*, **31**, 924–927.

7. NIOSH et al.

CHEW, D.L.E. (1988) Effective occupational safety activities: findings in three Asian developing countries. *International Labour Review*, **127**, 111–125.

COHEN, A. (1977) Factors in successful occupational safety programmes. *Journal of Safety Research*, **9**, 168–178.

COHEN, A. and CLEVELAND, R.J. (1983) Safety practices in record-holding plants. *Professional Safety*, 26–33.

COHEN, A., SMITH M., KENT, J. and ANGER, W. (1979) Self-protective measures against workplace hazards. *Journal of Safety Research*, **11**(3), 121–131.

SHANNON, H.S., MAYR, J. and HAINES, T. (in press) Overview of the relationship between organisational and workplace factors and injury rates. *Safety Science*, in press.

SIMONDS, R.H. and SHAFAI-SAHRAI, Y. (1977) Factors affecting injury frequency in 11 matched pairs of companies. *Journal of Safety Research*, **9**(3), 120–127.

8. Imperial College

DAWSON, S., POYNTER, P. and STEVENS, D. (1982) Strategies for controlling hazards at work. *Journal of Safety Research*, **13**(3), 95–112.

DAWSON, S., WILLMAN, P., BAMFORD, M. and CLINTON, A. (1988) *Safety at work: the limits of self-regulation*, Cambridge: Cambridge University Press.

9. Commercial auditors

DET NORSKE VERITAS (1994) *International Safety Rating System*, 6th, edn, Oslo: DNV.

GAUNT, D.L. (1989) The effect of the international safety rating system (ISRS) on organizational performance, Georgia: Center for Risk Management and Insurance Research, Georgia State University.

INDUSTRIAL ACCIDENT PREVENTION ASSOCIATION (1990) *The International Safety Rating System: an evaluation of effectiveness*, Toronto: IAPA.

PRISM (1994) *System and Reference Manual*, Washington, UK: AEA Consultancy Services.

10. Health and Safety Executive

HEALTH AND SAFETY EXECUTIVE, UK (1985) *Monitoring safety*, London: HMSO.

LINDSAY, F.D. (1992) Successful health and safety management: the contribution of management audit. *Safety Science*, **15**(3/4), 387–402.

11. Dutch Government

Handboek auditing (1988) Werkgroep audit, Directoraat-Generaal van de arbeid, Voorburg: Ministerie van Soziale Zaken en Werkgelegenheid.

Arbo- en verzuimbeleid (1993) Ministerie van Soziale Zaken en Werkegelegenheid, Den Haag: Staatsuitgeverij.

12. US management texts

BAILEY, C.W. and PETERSEN, D. (1989) Using perception surveys to assess safety system effectiveness. *Professional Safety*, **34**(2), 22–26.

HEINRICH, H.W., PETERSEN, D. and ROOS, N. (1980) *Industrial accident prevention*, 5th edn, New York: McGraw Hill.

PETERSEN, D. (1978) *Techniques of safety management*, 2nd edn, New York: McGraw Hill. (1993) Establishing 'safety culture' helps mitigate workplace dangers. *Occupational Health and Safety*, **62**(7), 20–24.

13. Chemical process industry

E&P FORUM (1994) *Guidelines for the Development and Application of Health Safety and Environmental Management Systems*, London: Exploration and Production Forum.

CENTER FOR CHEMICAL PROCESS SAFETY (1993) *Guidelines for auditing process safety management systems*, Washington: American Institute of Chemical Engineers.

14. US defence and aerospace/EG&G

JOHNSON, W.G. (1980) *MORT: safety assurance systems*, Chicago: National Safety Council of America.

15. Scandinavian participative management

CARTER, N. and MENCKEL, E. (1990) Group routines for improving accident prevention activities and accident statistics. *International Journal of Industrial Ergonomics*, **5**, 125–132.

EAKIN, J.M. (1992) Leaving it up to the workers themselves: a sociological perspective on the management of health and safety in small workplaces. *International Journal of Health Services*, **22**(4), 689–704.

ELDEN, M. (1983) Democratisation and participative research in developing local theory. *Journal of Occupational Behaviour*, **4**, 21–33.

HASLE, P. and OLSEN, P.B. (1995) The social conception of risk and development of motivation and commitment in health and safety work. Paper to the workshop Understanding the Work Environment, Swedish Institute for the Working Life, Stockholm, May 21–24.

KAMP, A. and RASMUSSEN, B.H. (1995) Working Environment and External Environment. Development of an Integrated Approach at Company Level. Papers to the workshop Understanding the Work Environment, Swedish Institute for the Working Life, Stockholm, May 21–24.

KJELLÉN, U. and BANERYD, K. (1983) Changing local health and safety practices at work within the explosives industry. *Ergonomics*, **26**(9), 863–877.

KOEFOED, L.B. (1995) Strategy for an interactive work environment and development of human resources. Papers to the workshop Understanding the Work Environment, Swedish Institute for the Working Life, Stockholm, May 21–24.

RASMUSSEN, B.H. and JENSEN, P.L. (1994) Working environment systems: trends, dilemmas, problems. In *Human factors in organisational design and management*, eds G.E. Bradley and H.W. Hendricks, pp. 43–48, Amsterdam: Elsevier.

SEPPÄLÄ, A. (1995) Developing MQW: from safety committees to daily actions. Papers to the workshop Understanding the Work Environment, Swedish Institute for the Working Life, Stockholm, May 21–24.

SIMARD, M. (1995) A systematic approach for developing participative management of safety at the shop-floor level. Papers to the workshop Understanding the Work Environment, Swedish Institute for the Working Life, Stockholm, May 21–24.

16. Sociological

DWYER, T. (1991) *Life and death at work: industrial accidents as a case of socially produced error*, New York: Plenum.

17. Confederation of British Industry

CONFEDERATION OF BRITISH INDUSTRY (1990) *Developing a safety culture: business for safety*, London: CBI.

18. INSAG

IAEA (1987) *OSART: guidelines for assessment*, Vienna: IAEA.
  (1988) *Basic safety principles for nuclear power plants*, 75-INSAG-3, Vienna: IAEA.
  (1991) *Safety culture: a report by the International Nuclear Safety Advisory Group*, Safety Series 75-INSAG-4, Vienna: IAEA.
  (1992) *ASCOT guidelines: guidelines for self-assessment of safety culture and for conducting a review: assessment of safety culture in organisations teams*, Vienna: IAEA.

19. Berlin

WILPERT, B. (1989) Conceptual blindspots in risk management: social dynamics, organisation and management. Paper to Second World Bank Workshop on Risk Management and Safety Control, Karlstad: Riskcentrum.
WILPERT, B. and KLUMB, P. (1993) Social dynamics, organisation and management factors contributing to system safety. In *Reliability and safety in hazardous work systems*, eds B. Wilpert and T.U. Qvale, Hove: Lawrence Erlbaum.

20. TNO

VAN STEEN, J.F.J. and KOEHORST, B.J.L. (1993) The SMART project: a framework and tools for addressing and improving management of safety. International process safety management Conference and Workshop, San Francisco, California, September.

21. Brookhaven/Pennsylvania State

HABER, S.B., METLAY, D.S. and CROUCH, D.A. (1990) Influence of organisational factors on safety. *Proceedings of the Human Factors Society*, 871–875.
JACOBS, R. and HABER, S. (1994) Organisational processes and nuclear power plant safety. *Reliability Engineering and System Safety*, **45**, 75–83.

22. Leiden/Manchester

REASON, J. (1993) Managing the management risk: new approaches to organisational safety. In *Reliability and safety in hazardous work systems*, eds B. Wilpert and T.U. Qvale, Hove: Lawrence Erlbaum.

WAGENAAR, W.A., HUDSON, P.T.W. and REASON, J.T. (1990) Cognitive failures and the cause of accidents. *Applied Cognitive Psychology*, **4**, 231–252.

WAGENAAR, W.A., SOUVERIJN, A.M. and HUDSON, P.T.W. (1993) Safety management in intensive care wards. In *Reliability and safety in hazardous work systems*, eds B. Wilpert and T.U. Qvale, Hove: Lawrence Erlbaum.

WAGENAAR, W.A., GROENEWEG, J., HUDSON, P.T.W. and REASON, J.T. (1994) Promoting safety in the oil industry. *Ergonomics*, **37**(12), 1999–2013.

23. Delft

HALE, A.R., GERLINGS, P.O., SWUSTE, P. and HEIMPLAETZER, P. (1991) Assessing and improving safety management systems, SPE 23241. In: *1st International Conference on Health, Safety and Environment*, pp. 382–388, Den Haag: Society of Petroleum Engineers.

HALE, A.R., HEMING, B., CARTHEY, J. and KIRWAN, B. (1997) Modelling of safety management systems. *Safety Science*, **26**(1–2), 121–140.

24. Safety climate scales

BROWN, R.L. and HOLMES, H. (1986) The use of factor-analytic procedure for assessing the validity of an employee safety climate model. *Accident Analysis and Prevention*, **18**(16), 455–470.

COOPER, M.D. and PHILLIPS, R.A. (1994) Validation of a safety climate measure. Paper to the Occupational Psychology Conference of the British Psychological Society, Leicester.

COX, S. and COX, T. (1991) The structure of employee attitudes to safety: a European example. *Work and Stress*, **5**, 93–106.

DEDOBBELEER, N. and BÉLAND, F. (1991) A safety climate measure for construction sites. *Journal of Safety Research*, **22**(2), 97–103.

NISKANEN, T. (1994) Safety climate in the road administration. *Safety Science*, **17**(4), 237–255.

ZOHAR, D. (1980) Safety climate in industrial organisation: theoretical and applied implications. *Journal of Applied Psychology*, **65**(1), 96–102.

25. Surrey/Liverpool

DONALD, I.J., CANTER, D.V., CHALK, J.R., HALE, A.R. and GERLINGS, P. (1991) Measuring safety culture and attitudes, SPE 23392: In: *1st International Conference on Health, Safety and Environment*, pp. 639–644, Den Haag: Society of Petroleum Engineers.

26. Kent

WEHRMEYER, W. (1994) Personal environmental attitudes and corporate environmental cultures: where do they come from and how do they mix? Paper presented to the Third Annual Business Strategy and the Environment Conference, Nottingham, September.

WEHRMEYER, W. and PARKER, K.T. (1995) Identification, analysis and relevance of environmental corporate cultures. *Business Strategy and the Environment*, **4**(3), 144–153.

27. VTT/Finnish nuclear industry

NORROS, L. and REIMAN, L. (1991) Coping with uncertainties and formation of safety culture in complex systems. In *Proceedings of the 9th NeTWork workshop on Safety Policy*, Bad Homburg NeTWork Workshop, Helsinki: VTT.

NORROS, L. and REIMAN, L. (1991) Uncertainties in the system as a challenge for NPP operators' expertise. In *Proceedings of the 9th NeTWork workshop on Safety Policy*, Bad Homburg NeTWork Workshop, Helsinki: VTT.

28. Risø, World Bank

RASMUSSEN, J. (1993) Learning from experience? How? Some research issues in industrial risk management. In *Reliability and safety in hazardous work systems*, eds B. Wilpert and T.U. Qvale, Hove: Lawrence Erlbaum.

  (1995) Converging paradigms of human sciences, Riskcentrum, University of Karlstad, Sweden, April 18.

RASMUSSEN, J. and BATSTONE, R. (1991) Safety control and risk management: toward improved low risk operation of high hazard systems. Findings from the world bank workshops, Riskcentrum, University of Karlstad, Sweden.

29. Berkeley

ROBERTS, K.H. (1989) New challenges in high reliability research: high reliability organisations. *Industrial Crisis Quarterly*, **3**, 111–125.

ROBERTS, K. (1990) Some characteristics of one type of high reliability in organisation. *Organisation Science*, **1**(2), 160–176.

ROCHLIN, G.I. (1988) Technology, hierarchy and organisational self-design: US naval flight operations as a case study. Paper to First World Bank Workshop on Risk Management and Safety Control, Washington.

  (1989) Informal organisational networking as a crisis-avoidance strategy: US naval flight operations as a case study. *Industrial Crisis Quarterly*, **3**(2), 159–176.

WEICK, K.E. (1987) Organisation culture as a source of high reliability. *California Management Review*, **29**(2), 112–127.

WIECK, K. (1989) Mental models of high reliability systems. *Industrial Crisis Quarterly*, **3**(2), 127–142.

30. Perrow

MESHKATI, N. (1989) Self-organisation, requisite variety and cultural environment: three links of a safety chain to harness complex technological systems. Paper to Second World Bank Workshop on Risk Management and Safety Control, Riskcentrum, University of Karlstad Sweden.

PERROW, C. (1984) *Normal accidents. Living with high risk technologies*, New York: Basic Books.

SAGAN, S.D. (1993) *Limits of safety*, Princeton: Princeton University Press.

31. (North) London

PIDGEON, N. (1991) Safety culture and risk management in organisations. *Journal of Cross Cultural Psychology*, **22**(1), 129–140.

PIDGEON, N.F. and O'LEARY, M. (1994) Organisational safety culture: implications for aviation maintenance. In *Aviation psychology in practice*, eds N. Johnston, N. McDonald and R. Fuller, Avebury Technical, Ashgate Publishing.

PIDGEON, N.F., TURNER, B.A., BLOCKLEY, D.I. and TOFT, B. (1991) Corporate safety culture: improving the management contribution to system reliability. In *Reliability '91*, ed. R.H. Mathews, Amsterdam: Elsevier.

TURNER, B. (1978) *Man-made disasters*, London: Wykeham Press.

TURNER, B.A. (1991) The development of safety culture. *Chemistry and Industry*, April 241–243.

32. NIA

ZWETSLOOT, G.I.J.M. (1994) The joint management of working conditions, environment and quality, Ph.D. thesis, Leiden University.

33. Standards

BS 7750 (1992) *Specification for environmental management systems*, London: British Standards Institution.

BS 8800 (1996) *Guide to health and safety management systems*, London: British Standards Institution.

PROPOSAL FOR NORWEGIAN STANDARD (1994) Occupational Health and Safety Management System-General Management Principles, Oslo: Norwegian Standards Institute.

3.4 Trondheim

FLAGSTAD, K.E. (1995) The functioning of the internal control reform: case studies in small and medium-sized enterprises. Doktor Ingeniøravhandling, Department of Industrial Management and Work Science, Norwegian Institute of Technology, Trondheim.

HOVDEN, J. and TINMANNSVIK, R.K. (1990) Internal control: a strategy for occupational safety and health: experiences from Norway. *Journal of Occupational Accidents*, **12**, 21–30.

PART FIVE

Safety interventions

There is a very large and diverse literature on interventions for improving workplace safety. Safety solutions vary enormously in scope and in type of approach taken. For example, solutions may involve simply the provision of safety information in passive form, such as posters and leaflets, or active forms, like safety lectures, or much more complex interventions such as those involving behaviour modification methods (Hale and Glendon, 1987). The main focus of workplace safety interventions is most often to reduce the risk of exposure to hazards and to prevent injury from occurring, although preventive efforts may also take a more protective focus with interventions to reduce the severity of injury or attempt to ameliorate the injury and shorten the time needed for rehabilitation after an injury has occurred. In Menckel's chapter in this part the implications of these different types of prevention are discussed in some detail.

In the past the emphasis has been on developing different types of solutions and there certainly have been trends in the types of solutions in use. These reflect the prevailing views about the causes of accidents and injury events. For example, earlier reports focused greatly on the need for barriers to prevent exposure to harmful energy of various types (Haddon, 1968); more recently there has been a focus on solutions that look at reducing hazards from the physical working environment (e.g. Sauter *et al.*, 1991) or that try to change unsafe behaviours such as various types of behavioural feedback programmes (McAfee and Winn, 1989).

Increasingly accident research is taking a systems approach to the causes of injury. In this approach the search for the causes of accidents emphasizes the multidimensional factors that contribute to the injury occurrence rather than simple linear causes (Laflamme, 1990). In particular, evidence from many sources is pointing in the direction of organizational factors playing a pivotal role in causes of accidents. For example, studies of fatal workplace accidents (Feyer and Williamson, 1991) and of less severe accidents (Reason, 1990; Runciman *et al.*, 1993) have shown that pre-existing factors, often relating to the way the work is done and organized, lay at the genesis of these accidents (see Part Four for a further discussion of this area of research).

The diversity of types of interventions means that it is probably not very useful to talk about safety interventions as a single category. We should instead be looking at what types of interventions need to be developed based on what works. The natural consequence of this change in approach has been the realization that not all preventive strategies and measures will have the same impact on improving work safety. This has led to the further realization recently that although there has been a wealth of ways of approaching injury prevention there has been little attempt to try to evaluate their success. In fact well-designed evaluations of safety intervention programmes have been very few and far between.

One of the most recent attempts to redress this was a review of safety interventions by Guastello (1993). Using standard measures of effect size for each evaluation study, Guastello was able to compare the effectiveness of different types of interventions for reducing accidents. His findings revealed that comprehensive ergonomics programmes and interventions based on behaviour modification had quite impressive effects on reducing accidents. On the other hand, as a group, personnel selection techniques had very little influence on accident rates.

This work is an important start in trying to work through the current literature on interventions to determine what we know and what we need to know. It is, however, as Guastello himself says, only a start. While there are some signals emerging right now from the Guastello analysis about what type of interventions might be most successful, there is still a great deal of uncertainty about why some types of interventions are more successful than others and why particular types are successful sometimes but not always. As a case in point, in Guastello's review there was considerable variation of effect sizes between studies using the same intervention, for example the two evaluations of poster campaigns had effect sizes of −6% to 33%.

The chapter by Shannon in this part describes the results of a survey of the organizational characteristics, particularly with respect to occupational health and safety, in a large sample of Canadian industry. The results indicate the importance of focusing on organizational factors as the most productive candidates for safety interventions. More specifically, the results show which organizational characteristics appear to be important for determining injury rates and which ones do not. For example, an active role of top management in safety was significantly associated with lower injury rates whereas the existence of written safety rules was not. This is an important study because it is one of the few that have attempted to determine empirically how important organizational variables actually are and which are the most important.

There are many factors that are common to the process of intervening that may play a critical role in determining how successful a particular type of intervention actually is. Saari's chapter in this part discusses the importance of taking cultural differences into consideration in all phases of a particular intervention. He cites a number of examples of intervention failures where the failure was most likely due to cultural factors. In some instances these were national or ethnic cultural factors, but in others they were the difference between good and poor companies. This means that interventions cannot just be taken off the shelf and be expected to be successful. Saari argues that their impact needs to be evaluated at all phases of the intervention, including acceptance, planning and implementing.

The chapter by Shaw and Blewitt in this part also takes up the call for more systematic and intensive evaluation of interventions. They argue that this is the only way that we can really be effective in intervening and also the only way that our interventions will really improve. They make a strong case for including more than simple accident rates in evaluations. Accident rates are, as Shaw and Blewitt point out, the last step in the accident/injury process. There is a wide range of alternative indicators, some of which are objective, but many of which are necessarily subjective, and these need to be used to reflect the progress of the intervention.

In Menckel's chapter in this part, the concept of cooperation is added to the list of prerequisites for successful interventions. As well as prevention and intervention, Menkel puts forward a model of injury prevention where cooperation is conceived of as a form of glue which brings together the researchers, the practitioners and the shopfloor personnel, including workers and supervisors. Menckel maintains that activity to prevent injury requires collaboration between researchers and practitioners and in many cases teamwork involving the many stakeholders in workplaces. This, she maintains, interacts with the nature of the intervention itself to determine how successful any intervention will be in the workplace. It is very likely that cultural features will play a role in the level and nature of cooperation achieved in any intervention and that the concept of cooperation is one of a number of factors that make workplaces differ and interventions succeed or fail. The path to developing successful safety interventions for all workplaces depends on gaining a better understanding of how these influences affect intervention programs.

References

FEYER, A.-M. and WILLIAMSON, A.M. (1991) A classification system for causes of occupational accidents for use in preventive strategies. *Scandinavian Journal of Work and Environmental Health*, **17**, 302–311.

GUASTELLO, S.J. (1993) Do we really know how well our occupational accident prevention programs work? *Safety Science*, **16**, 445–463.

HADDON, W. (1968) The changing approach to the epidemiology prevention, and amelioration of trauma: the transition to approaches etiologically rather than descriptively based. *American Journal of Public Health*, **58**(8), 1431–1438.

HALE, A.R. and GLENDON, A.I. (1987) *Individual behaviour in the control of danger*, Amsterdam: Elsevier.

LAFLAMME, L. (1990) A better understanding of occupational accident genesis to improve safety in the workplace. *Journal of Occupational Accidents*, **12**, 155–165.

MCAFEE, R.B. and WINN, A.R. (1989) The use of incentives/feedback to enhance workplace safety: a critique of the literature. *Journal of Safety Research*, **20**, 7–19.

REASON, J. (1990) The contribution of latent failures to the breakdown of complex systems. *Philosophical Transactions of the Royal Society of London B*, **327**, 475–484.

RUNCIMAN, W.B., SELLEN, A., WEBB, R.K., WILLIAMSON, J.A., CURRIE, M., MORGAN, C. and RUSSELL, W.J. (1993) Errors, incidents and accidents in anaesthetic practice. *Anaesthesia and Intensive Care*, **21**, 506–519.

SAUTER, S.L., DAINOFF, M. and SMITH, M. (1991) *Promoting health and productivity in the computerized office. Models of successful ergonomic interventions*, London: Taylor & Francis.

CHAPTER TWELVE

Workplace organizational factors and occupational accidents

HARRY S. SHANNON

12.1 INTRODUCTION

In 1994, Nancy Krieger from the US wrote a paper entitled *Epidemiology and the Web of Causation: Has Anyone Seen the Spider?* (Krieger, 1994). In it Krieger described the implicit reliance in much of epidemiological theory and practice on what she called the 'framework of biomedical individualism' to guide the choice of factors incorporated in the thinking of epidemiologists and hence in the study designs they use. Krieger argued that it was more important to question the whereabouts of the putative spider, who was creating and spinning the web. She was pointing out that focusing on the micro-level misses the bigger picture in which the societal factors influence what happens at the individual level. This has important implications for those in the field of occupational safety and expresses the rationale for the studies to be described. They will show correlations between what happens at the workplace level and injury rates within those worksites.

In the late 1970s a new Occupational Health and Safety Act came into effect in the province of Ontario (Canada). Since that time, the rate of fatal accidents in Ontario, and indeed the rest of Canada, has shown a decline (Figure 12.1). While some of this may be due to a change in the nature of work over this time (fewer people in the risky industries of manufacturing and construction, for example), analyses have also been done showing declines by industry. This drop is notable and welcome. A decline has also been shown for all compensation claims. However, there was essentially no reduction in lost-time injury rates during the 1980s (Figure 12.2), although there has been a decline since then, which is probably mainly due to the recession. Given the intensive efforts in health and safety over this period of time these results must be considered as disappointing. More importantly, they raise questions about the focus of our preventive efforts.

Many of these efforts are aimed at the micro-level, for example attempting to screen workers for such things as alcohol or drug use. However, a review by Guastello (1993) showed a minimal, if any, effect of this approach. On the other hand, rather larger effects were found for programmes at the mezzo- and macro-

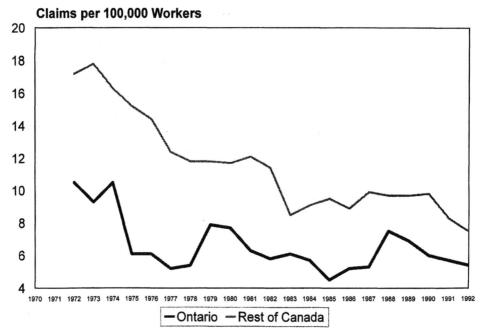

Figure 12.1   Fatality rates. Source: Human Resources Development, Canada.

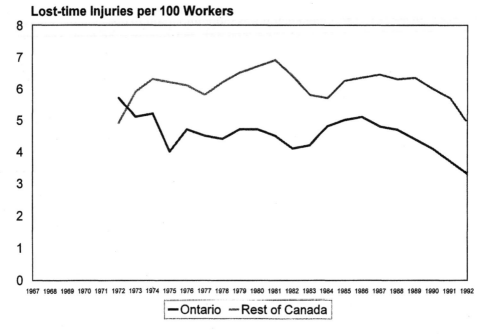

Figure 12.2   Lost-time injury rates. Source: Human Resources Development, Canada.

level (especially the workplace). In addition, there are some workplaces that strongly emphasize safety and have consistently good safety records, as much as an order of magnitude better than the average for their industries. It has been suggested that the general philosophy and culture of these workplaces is responsible for this. These observations emphasize the potential role of workplace organizational factors in safety, and in particular which factors may be relevant.

This point is not new – it is well known to those who study organizations. As Katz and Kahn wrote 30 years ago '. . . attempts to change organizations by changing individuals have a long history of theoretical and practical failure . . . Its essential weakness can be labelled the *psychological fallacy*, the concentration on individuals without regard to the situational factors that shape their behaviours . . . it is a great over-simplification' (Katz and Kahn, 1966).

12.2 STUDY OF HEALTH AND SAFETY APPROACHES IN THE WORKPLACE

Several years ago, some colleagues and the author at McMaster University in Ontario started a study to address this topic (Shannon *et al.*, 1996). The first step involved developing a conceptual model of organizational factors in relation to health and safety (Figure 12.3).

12.2.1 Conceptual model

On the left-hand side of the model (Figure 12.3) are external factors such as the regulatory environment or economic pressures like 'global competition.' Since these are (in general) outside the control of the workplace we leave them aside. (Certainly a look at the 'really big picture' would consider them to be of great relevance.) The middle box on the left shows that managers and workers bring to the workplace

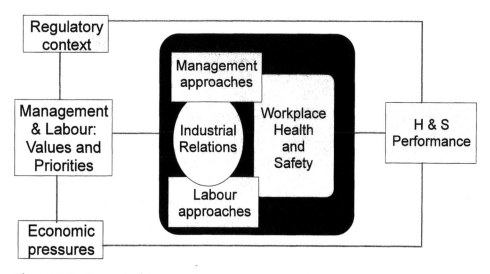

Figure 12.3 Conceptual framework.

certain values and priorities. In addition, inherent in the terms 'manager' and 'worker' are notions of authority and power (and there are some legal implications as well). Within the workplace itself, shown by the central area, we have factors related to management. For example, they are responsible for determining the socio-technical context, the general organizational culture and so on. At the same time, labour may or may not be organized, it may be more or less active, etc. Labour and management interact through industrial relations. As well as these general issues of labour and management behaviour, we have their specific approaches to health and safety. Finally, these factors all lead, directly or indirectly, to some level of performance in health and safety.

It is noteworthy that the model does not include directional arrows in this diagram. This is because one could easily conceive of a relationship in either direction between any two boxes. For example, poor health and safety performance will almost certainly feed back to influence the priorities of managers and workers. Nevertheless, the general understanding is that the direction of effect is from left to right.

12.2.2 Methodology

The target population of the study was companies in the manufacturing and retail sectors in Ontario. (Our funders, the Industrial Accident Prevention Association, have a mandate to deliver health and safety programmes to these sectors.) The first phase was a survey of labour and management representatives in the workplaces. The second phase was a qualitative study involving in-depth interviews with several people at each of a small number of companies. This second phase will not be explicitly discussed, although its findings were of great help in interpreting the survey results.

Based in part on the conceptual model, it was decided to ask questions about the following areas:

- organizational structure and philosophy;
- organizational philosophy on OHS;
- labour markets and unions;
- internal responsibility system;
- organizational demographics;
- risk and physical conditions; and
- financial performance and profitability.

There did not appear to be a questionnaire or questionnaires that covered this broad range of topics. A survey in Ontario on joint health and safety committees provided a number of questions. In addition, based on an understanding of the topics and theoretical considerations (for example, ideas from the management literature), items of interest were listed and questions were written designed to capture the relevant information. Separate questionnaires were aimed at management and labour, although some questions were deliberately repeated on the two questionnaires to look for similarities or differences in the responses. The management questionnaire was divided into three booklets: one each for the senior manager, human resources (or personnel) manager, and the co-chair of the health and safety committee. (In some smaller companies, the same person might fill more than one of these roles.) The questionnaire was pilot-tested on a small number of companies and we incorporated the feedback in revisions.

Workers' compensation data were used to help select the workplaces in the sample. Eight rate groups were chosen, i.e. groups of companies doing similar types of work and so-called because they form the basis for the *rate* of workers' compensation premium. The six manufacturing groups were metal articles, plastic articles, automobile assembly, grain products, textiles and printing. For the retail sector, the two categories were food retail and other retail, that is general merchandise. Attention was restricted to companies with more than 50 employees (more than 20 in retail operations) and they were stratified by lost-time frequency rate (LTFR) and workplace size. The size stratification was a division into companies with 50–99 employees and those with 100 or more. LTFRs were considered high if they were more than 50% above the rate group average and low if they were less than 50% of the average. LTFRs in between were labelled medium. We generally over-sampled companies in the low and high strata. Overall, there were slightly more than 1000 companies in the sample. Intensive efforts were made to achieve a good response rate, using reminder letters and several telephone calls if necessary. Overall, the response rate (at least one booklet returned) was 52%. It was a little higher for the manufacturing workplaces, whose results will be used. This is a remarkably high response for a survey of this nature.

Some workplaces reported verbally that the questionnaire was too long and they did not have time to complete it. Given this, we selected a subset of questions that were asked in a telephone survey of those who had not replied to the written questionnaire. We contacted 200 workplaces (randomly chosen from the non-responders) and were able to carry out the interview in 75%. (Among worker representatives, this figure was smaller, around 50%, but we believe this was because of difficulties in contacting workers on the job.) The distributions of replies to questions in the telephone interview were very similar to those obtained from the mail survey. In addition, the response rates to the mail survey were very similar in companies with different safety performance, that is with high, medium or low lost-time frequency rates. We thus concluded that comparisons between workplaces with different claim rates were valid and not biased by response patterns.

Following preliminary data examination, univariable analyses were conducted by comparing variables in the three LTFR categories. In a final series of analyses we examined variables simultaneously in multiple regressions. However, these regressions were of limited value because even given a response from a company, some data needed for the regression was not provided.

12.2.3 Results

Many variables were considered so in interpreting the results we looked for similar patterns in groups of variables as well as a consistent trend across the three LTFR categories. Companies with older workers, whose workers who had longer seniority and with low turnover tended to have lower LTFRs. These characteristics are, of course, correlated, so one cannot tell which, if any, may lead to lower accident rates. Indeed, it may be that companies that treat their workers well and retain them are also better at maintaining safer workplaces.

Lower grievance rates and better labour relations (at least as perceived by management) were related to LTFR, as were encouragement by management of long-term career commitment and provision of a long-term disability plan. These variables show managements who demonstrate a concern for the workforce by concrete actions. Both greater worker participation and lower expectations that workers will simply follow management instructions were related to lower LTFR. This suggests that 'empowerment' is important. Interestingly, the existence of a written health and safety policy was not related to lower rates, nor was the status of the health and safety coordinator (where one was present). Several factors that were related included defining health and safety in every manager's job description, the importance of health and safety performance in managers' annual appraisals and attendance by the senior manager at health and safety meetings. These suggest that concrete actions, rather than verbal or implicit commitment, are important for good health and safety.

Few features of the health and safety committee were related to LTFR. The main one that was related to claim rates was concerned with its problem-solving style. When labour or management more often threatened to go outside the committee, the LTFRs were higher. This suggests that making efforts to solve disagreements internally is a characteristic of more effective committees.

12.3 OVERVIEW OF STUDIES ON WORKPLACE ORGANIZATION AND SAFETY

One limitation of this type of study is that with many questions being asked, some of the organizational variables will be significantly related to the claim rates by chance alone, even if there is truly no relationship. The standard scientific approach to this is to demand replication of the results by conducting further research, etc. In this instance, several reports became available during the conduct of our research. We have therefore synthesized results from the studies, thus in effect attempting to replicate our own data (Shannon *et al.*, 1997). The aim was to look for consistency from study to study, that is those variables that were consistently related to accident rates from study to study; other variables might only be significant in a single study. One would thus have much greater confidence that the consistent variables were truly related to injury rates and not significant simply as a result of chance.

12.3.1 Methods

We looked for studies compiled from primary data and using a quantitative methodology. They had to use an outcome (not process) measure to relate to organizational variables and had to make a comparison between companies with different accident rates. At least 20 organizations had to be included. Searches were made of a number of computer bibliographic databases such as NIOSHTIC and CIS-ILO. In the papers identified, we looked for variables reported in at least two studies. Among these, we checked whether they were significantly related (within each study) to the injury rate, taking note of the direction of the relationship. Variables were classified under four broad headings: joint health and safety committees, management style and culture, organizational philosophy regarding health and safety, and other. A variable was defined as consistently related to injury rates if (a) the relationship was statistically significant in one direction in at least two-thirds of the studies in which it was examined and (b) there was no study in which it was significant in the opposite direction. A variable was labelled as contradictory if there was at least one study in each direction in which a relationship was significant.

12.3.2 Results

A number of variables met the criteria for consistency. These included empowerment of the workforce, an active role of top management and greater seniority of the workforce. Among the variables with contradictory relationships were the level and use of discipline for safety violations and the profitability of the company. (The contradiction in the latter occurred only between subgroups within a single study.) In addition, some variables that were (perhaps surprisingly) *not* consistently related to injury rates were noted. These included the existence of written safety rules and whether a senior or middle management representative sat on the joint health and safety committee. Several of these studies showed how much of the variability between accident rates was accounted for by the variables they had studied. The proportion varied from 13% to 47%. Given the potential errors in measurement of the data, these values can be considered comparatively high. Of course, they simply refer to correlations rather than necessarily causation.

12.4 CONCLUSIONS

It now appears to be important to apply interventions in the workplace based on these results to determine whether appropriate changes to the workplace organization do indeed create better health and safety conditions. This is particularly relevant given the changing nature of work and the workforce in most industrialized countries. For example, if there are increased pressures at work, the pace of work rises and control over one's job may fall. These factors have been linked to general health outcomes and have now been shown to be related to safety in particular.

Hardly an organization exists that is not being re-engineered, restructured or changed in some way, and there are analyses of historical data suggesting that during periods of rapid changes of any kind (technological, structural and so on) adverse effects on accidents can be expected (Saari, 1982). It is vital to ensure that the health and safety effects on the workforce are among the factors considered by those involved in creating the change. To do this we must keep in mind the big picture and, as Krieger urges us, not lose sight of the spider.

ACKNOWLEDGEMENTS

I thank my colleagues at McMaster University with whom I conducted the research described in this chapter: Ted Haines, Wayne Lewchuk, Janet Mayr, Lea Anne Moran, Jack Richardson, Dave Verma and Vivienne Walters. The Industrial Accident Prevention Association of Ontario and the Institute for Work & Health provided funds for and supported the projects.

References

GUASTELLO, S.J. (1993) Do we really know how well our occupational accident prevention programs work? *Safety Science*, **16**, 445–463.

KATZ, D. and KAHN, R.L. (1966) *The social psychology of organizations*, New York: Wiley.

KRIEGER, N. (1994) Epidemiology and the web of causation: has anyone seen the spider? *Society of Scientific Medicine*, **39**, 887–903.

SAARI, J. (1982) Accidents and progress of technology in Finnish industry. *Journal of Occupational Accidents*, **4**, 133–144.

SHANNON, H.S., WALTERS, V., LEWCHUK, W., RICHARDSON, R.J., MORAN, L.A., HAINES, T. and VERMA, D.K. (1966) Workplace organizational correlates of lost-time accident rates in manufacturing. *American Journal of Industrial Medicine*, **29**, 258–268.

SHANNON, H.S., MAYR, J. and HAINES, T. (1997) Overview of the relationship between organizational and workplace factors and injury rates. *Safety Science*, **26**(3), 201–217.

CHAPTER THIRTEEN

Safety interventions: international perspectives

JORMA SAARI

13.1 INTRODUCTION

International communication has become much easier and it continues to become faster and accessible for many more people. The Internet and other new means of communication will make it very easy to exchange news and to transfer experiences from one country to another. Multinational companies and consulting firms use international benchmarking in an endeavour towards excellence in management and production. Best practices are sought internationally also in occupational health and safety. It is tempting to rely on a formula that has some demonstrated benefits in another country, industrial sector or company. Mistakenly, we transfer formulas from country to country without considering any potential barriers.

International competition induces such efforts. Everybody tries to learn the competitors' secrets and to know the reasons for their success. How well does the transfer of successful formulas from one country to another work? Is it really possible to copy other companies' successes? This chapter highlights some of the issues involved. It is primarily based on the author's personal experience and his observations from several major intervention studies he has been involved with. It is also based on his research experiences from two continents, especially from Finland and Canada.

Most recently, researchers have started to become aware of the risks of copying ideas from country to country. However, the topic is not totally new as Suchman's article from 1965 demonstrates (Suchman, 1965). Janssens *et al.* (1995) recently studied a US-based multinational company running a similar safety policy in all its plants. A questionnaire study in three plants in three countries, the USA, France and Argentina, showed quite different responses to the corporation's safety policy. Despite the same policy, the impact on the employees' perceived safety was quite different in each plant. Janssens *et al.* (1995) concluded that various factors in the style of leadership, such as individualism/collectivism, authoritarian/paternalistic management style and autocratic/participative decision-making, moderated the outcomes of the corporation-wide safety policy. Thus, the same policy can lead to quite different perceptions among employees about management's safety concerns.

Liker *et al.* (1989) found that participatory ergonomics in Japan and the US actually involved totally different applications. These applications were also moderated by the differences in the management style and in culture. Liker *et al.* (1989) compared results from two corporations using a participatory approach in ergonomics. The applications were quite different but the results were good in both cases. This is an example of taking a concept and modifying it to fit the local conditions. This participatory approach is an example of a safety technique that is currently attracting a lot of interest internationally. It has had quite obvious success in the Scandinavian countries. In North America interest has been accelerating in the effectiveness of health and safety committees. However, the positive results have not necessarily materialized in the same way as in Scandinavia (Sheehy and Chapman, 1987).

In Ontario, a bipartite Workplace Health and Safety Agency was formed in 1991. It duplicated a model which has been functioning with good results in several Scandinavian countries for decades. However, the model failed and the new provincial government folded the agency into the Workers' Compensation Board in 1996. The most important failure was that the agency did not maintain a reasonable balance of power between management and labour. Its work was slowed down considerably because the two partners could not reach consensus on many critical issues. Bipartitism was adopted in a new cultural environment in which confrontation and adversarial relations between unions and management were a tradition. Neither side was properly organized; unions represent only a fraction of all workers but claim full representation and companies are not systematically organized into management associations.

Perhaps the most critical reason for the failure was the rejection of a neutralizing factor, namely researchers and professionals. The application disqualified researchers and other experts as agenda setters. The process became purely political. In fact, the two sides started negotiating about issues well beyond their control. They bargained about how much risk would be tolerated, but not what the risk in each situation is. A well-functioning bipartite process is based on facts found out by researchers and experts. The labour market partners then decide what is a tolerable amount of risk. If the factual basis is neglected, the process loses its trustworthiness. This is what happened in Ontario.

In the Ontario example a good excuse for not having research properly included in the process was the lack of OHS research in the province. The neighbouring province, Quebec, has used a bipartite approach without a major failure. When bipartitism was seriously adopted in Quebec, a major research investment was also made, thereby introducing a neutralizing factual knowledge base. In Ontario, the agency created a number of differences of opinion and it induced a lot of dissatisfaction at work places, especially among management. This situation then gave to the new government a legitimate reason to have the agency's functions reviewed and to ultimately fold it into the Workers' Compensation Board in a manner which brings bipartitism back to square one.

This single case illustrates well that a successful formula in one place may not work elsewhere unless the causes and conditions of success are properly understood. The case does not prove that bipartitism cannot work in Ontario. It only tells that any successful formula needs to be well understood and, if necessary, adjusted to a new setting before implementation.

13.2 AN INTERVENTION MODEL

Several large intervention studies have been carried out by the author over the years. This experience is the basis for a model to discuss various potential barriers around transferring safety programmes from one setting to another. Figure 13.1 indicates some of the steps of the process that are critical to a successful intervention and the barriers that are discussed in this chapter. The conditions and factors around

1. Acceptance
- mental model of safety
- real vs apparent acceptance

2. Planning
- traditions
- safety culture

3. Implementation
- good companies implement everything better

4. Change and outcome effects
- maturing with time
- what are the real outcome criteria?

Figure 13.1   Some critical steps in a safety intervention and some barriers the article discusses.

these steps need to be known to avoid failure when using the same intervention in a different situation. This chapter does not try to be exhaustive; it describes cases from the author's experience.

13.2.1 Acceptance

People are most likely to accept safety programmes which match with their own mental models about safety. When we arranged OHS courses in Tanzania in the 1970s, many participants 'knew' that milk prevents silicosis. A programme providing milk to workers exposed to silica dust would have been very acceptable there, while in Europe practically nobody would have accepted such a 'safety' programme. Most likely, a milk programme would have shown some good results in Tanzania by improving the participants' overall health through better nutrition.

A person with a long career in line management more readily accepts safety programmes that resemble normal management procedures. A safety professional who knows about accident causation understands that managing production and managing safety are two different tasks and the same techniques do not work for both. In production, a manager gets more products with measures such as working overtime, using better machines, using incentives, rewards, etc. Everything is fairly predictable. Safety is different. A mere order does not suffice and money cannot buy safety. The mechanisms of accidents are often subtle and not well understood. Production mechanism is visible and designed by man. Accidents are under the influence of several uncontrolled factors and are not designed by man.

One example of the difference between safety and production management is quality errors in production. It is common that the quality of products is monitored through statistical process control. If errors appear, the causes are laid down and corresponding measures taken. The same approach should work for safety also.

Accident investigations require proper cultural support

Learning from accidents and other unwanted outcomes is the main avenue for improvements in occupational safety and ergonomics according to textbooks. Heinrich positioned accident investigations centrally in his prevention strategy (Heinrich *et al.*, 1980). In the late 1970s many people in Finland claimed that we need a good method for supporting accident investigations in workplaces.

Learning from accidents has quite obviously had good results even in the past. A Viking returning safely to home had an obligation to immediately build a new ship, taking into account the experiences of the most recent ship. In this way Vikings created a shipbuilding technology that modern technology surpassed only this century. Aviation is a modern example of investigating all incidents carefully. The idea of improving accident investigations sounds good.

The association representing Finnish occupational accident insurance companies thought that a good investigation method was needed urgently. The association named a group consisting of representatives from unions, employers' associations and insurance companies. The author became a member of this group and it was felt that a method to investigate accidents would really be a step forward. The group prepared a method that was actually a set of methods, starting from a very simple one and ending in a rather complicated model (Tuominen and Saari, 1982; Saari *et al.*, 1983). The association published a booklet, a short version of the booklet and a video about the investigation model. Several short courses were arranged in different towns and hundreds of people from workplaces participated. Many companies invited the group to their internal training sessions.

A couple of years later the group carried out a questionnaire study about the impact the model and the associated training sessions had on accident investigations in workplaces. The result was disappointing: the impact had been quite modest. An explanation for the lack of impact came from a company shortly after the study. An older supervisor pointed out 'If we really investigated all accidents as you say, we would soon be in trouble. It happens from time to time that preventive measures are not implemented even if identified. If we documented everything and another similar accident happened, we would be in real trouble. So it is better not to investigate at all.'

The regulations and public opinion did not support thorough investigations. If the management, or public opinion, had demanded investigations, the model would have been utilized. In another country where such conditions existed accident investigations might have been a good element of safety programmes. In those days in Finland, they were not.

Aviation, nuclear power production and the chemical industry are examples of industrial sectors that have developed a culture that promotes learning from accidents. Sea transportation is an example of the opposite culture. According to Linstone and Mitroff (1994), 15% of oil tankers have some kind of collision each year. Accident investigations do not lead to better sea safety because the economic conditions in the industry do not support safety and, probably, because of widespread fatalism within the industry. For example, Icelandic seamen are more likely to have a fatal accident than their fellow countrymen, and not only at sea (Rafnsson, 1993). According to Rafnsson, seamen are modified by their occupation toward hazardous behaviours or risky lifestyles.

Very large ferries, which carry cars and passengers, travel between Finland, Sweden and Estonia. In 1993 1000 passengers had to be evacuated from a ferry that touched the ground. Fortunately, the incident happened in daylight, in good weather conditions, and nobody was hurt. The shipping line did not launch immediate corrective measures. On the contrary, the management disparaged the whole incident and claimed that nothing serious had happened or could happen because of modern navigation technology. Shortly after this incident, the author wrote an article (Saari, 1995) stating that 'the described event demonstrated the possibility of a major accident'. Only a few months later, the *Estonia* travelling from Estonia to Sweden, sank in bad weather and about 900 passengers died in the accident.

The effect of a poster campaign

A study was carried out on the effects of a poster campaign on safety at a shipyard (Saarela *et al.*, 1989). The campaign was designed to meet all the criteria of a good effective campaign proposed in the literature (Hale and Glendon, 1987). The campaign was run on two ships, which were the first ships of the respective production series. The second ships in the series served as comparison ships.

Randomly chosen employees from the campaign ships were interviewed and it was found that they had seen the campaign posters. They could recall the messages amazingly well. During the interviews, however, it was observed that the messages had a very limited behaviour effect. The injury rates of campaign ships did not reflect any effect. The conclusion was that people had noticed the campaigns quite well, but as accident prevention they were useless. However, when asking peoples' opinions about the campaigns, they were always very positive: typical responses were 'It is really good to have safety campaigns at a workplace' and 'The campaigns reflect the management's safety concerns in a positive way'. During the following years the author's team was invited to the shipyard twice to conduct a major research project there. These invitations were based on the positive experiences from the campaign study.

In this situation the safety measure had no effect on safety. However, there was a wide acceptance of the safety measure. The safety campaign matched well with peoples' mental model of safety. If a careful evaluation of accidents had not been made, the campaign would have been promoted as an effective programme.

There is evidence that safety campaigns can have positive effects on behaviour (Laner and Sell, 1960) and on accidents (Kaestner et al., 1967). It may be, however, that the effectiveness of a safety campaign depends on timing. Safety campaigns aim at increased awareness. When the messages are new, they do increase awareness. Probably, the campaign described above just more or less repeated what was already known. The lack of awareness was not the real problem; other factors reinforcing unsafe behaviour and unsafe conditions caused the problem.

This example also emphasizes the importance of using multiple criteria when evaluating outcomes. The value of a poster campaign may not have a direct effect on injuries. Its most valuable effect may be in preparing the ground for other safety actions.

13.2.2 Planning phase

Even if there is sufficient acceptance of a safety programme, it may face severe difficulties. Apparent acceptance, just as in the case of accident investigations, may be very misleading. However, it is sometimes possible to make the necessary corrections during the planning phase.

An intervention study was carried out in a customs office in western Finland in 1992–95 (Kivistö, 1995). Customs is a centuries old organization that has always functioned in a paramilitary fashion; senior officers plan the operations and make the decisions, and frontline officers obey the orders they receive. However, this model does not work very well in the modern world. International crime continually develops new methods for smuggling and for other criminal acts. The old paramilitary model does not respond quickly enough. An organizational model imitating the human nervous system would be more appropriate. In critical situations action should come first and acceptance later. A more participatory type of management is therefore desirable. The frontline customs officers should be involved in the planning of various operations and other tasks. This would relieve unnecessary stress, even if the problems were properly taken care of by management. For example, the special investigation team in the customs office studied has to enter foreign ships for very thorough investigations. The personnel were worried about the risk of infectious diseases and wondered if they had been vaccinated against all relevant diseases. The management, together with the occupational health services, decided about the vaccinations without consultation with the personnel. One can believe that the decision was based on appropriate expert knowledge. However, because enough information was not disclosed, the officers became concerned.

The management of this customs office had realized the need for change. The author and his team happened to approach the customs office at the right time and were accepted because of the participatory techniques they had developed earlier (Saari, 1992). The management made the decision. The union official was involved but said nothing as he was used to obeying superiors' orders. The team had an apparent acceptance from the organization.

The purpose of the study was to initiate an experiment with participatory teams in which the frontline customs officers, together with their supervisors and middle managers, would set objectives, solve problems and measure performance. These objectives were related especially to safety and well-being but also to efficiency and quality. The strategy chosen was to start from easy objectives and, after having succeeded in these, continue to more demanding objectives.

It became obvious from the beginning that the teams could not function before an anonymous interview study was done. Anonymity was the only way officers were ready to share their opinions with their superiors. An open discussion was clearly rejected at the beginning.

When the interview study was complete, after several months, the time was thought to be suitable for returning to the original plan. A one-day meeting was arranged at which the upper management, middle management and frontline officers discussed their overall goals and objectives. For them, this was the first seminar of this type, ever.

There was a very heavy burden of tradition inhibiting discussions in participatory teams. Even if the upper management strongly encouraged this approach, some of the teams needed months to get organized. One of the teams never functioned really well. Participatory development teams were a good idea but the team members carried a long tradition of one-way communication. Because this was an experiment in one customs office, the team members were not sure if this really was the model to the future. Uncertainty made them careful in expressing opinions and thoughts.

The customs office had two very different functions: (1) the control of passenger and goods traffic, and (2) a purely administrative function for levying duties and taxes. The first function operated in the paramilitary fashion. The duty and tax department was located in a separate building and their organizational culture was similar to any other governmental office. There were also participatory teams in the duty and tax department. These teams started working quite quickly. Their supervisors were 'old-style' customs officers. Even this did not deter them from the fast initialization of the team process.

Another unanticipated roadblock was met in this study. The basic strategy was based on a slow, gradual move to the new direction. Therefore the participatory development of operations started with easily accepted objectives, such as self-defence skills, and, after having gained some satisfaction from attainment, moved to the more difficult and controversial objectives, such as the distribution of working hours between mornings and evenings; there was a peak of passenger traffic around 8 a.m. and around 8 p.m. that made the planning of shifts a problem.

Previous experience and other research has shown that early success can make people more motivated to a new programme (LaBarr, 1995). Attacking less difficult problems first can build the confidence and satisfaction that bigger changes require.

This strategy failed and no way was found to correct it. Two external changes contributed to the failure: the collapse of Soviet Union and Finland's speculated membership of the European Union (EU). These events induced a major threat to the customs office as it was located in the western part of Finland. It was thought that Finland's membership of the EU would reduce the need of customs services in the west and the collapse of Soviet Union would create a bigger goods and passenger traffic across the eastern border. The customs office was therefore losing importance and people perceived a major threat to their job security. For these reasons, everybody was anxious to get their job security ensured. They did not have the mental tolerance for some less important changes in the workplace. Therefore some more demanding objectives had to be accepted from the beginning and this quite obviously slowed down progress. In the end, Finland became an EU member. However, it also became clear that the membership had less effect on the customs work than anticipated during the study and, due to the relief of mental stress, the participatory development then progressed quickly. This case shows how local traditions and external changes affecting the organization can force a modifcation of plans.

Example of failure in an attempt to measure safety climate

When in Montreal at McGill University, the author started a study on the implementation strategies of WHMIS (Workplace Hazardous Materials Information System). This is a legislated safety programme in Canada that requires every company to have their employees trained in the handling of hazardous materials, to have all the containers of hazardous materials labelled and to have an on-site collection of materials safety data sheets (MSDS) available for the employees. Companies used different training strategies: some had all employees trained by external experts, some had trainers trained and some companies did not comply with the regulations on schedule. This provided an excellent opportunity to evaluate the different training strategies (Saari et al., 1991; Saari et al., 1994). The study was carried out in the sector manufacturing transportation equipment and machines. This is a small manufacturing sector employing approximately 50000 people and serviced by a bipartite safety association (SATEMM-ASFETM). With the help of the safety association, we obtained for the study a sample of 80 companies situated in the province of Quebec, Canada. This is a large province and some companies were hours away from Montreal. In each company several aspects of the company itself, their WHMIS training results and the WHMIS system were evaluated.

It was thought that the company's safety climate would be a moderating factor. A company with a good safety climate would possibly have better training results, better labelled containers and better organized collection of MSDSs. To measure the safety climate a short questionnaire, based on Dedobbeleer's work in the American construction industry (Dedobbeleer and German, 1987), was designed. Some experience from a fairly big intervention study conducted in Finland a few years before this study was also available (Saari, 1986; Seppälä et al., 1987). Safety climate was one of the dependent variables in the study and it seemed to measure the improvement of companies' safety quite well.

A shortened questionnaire was designed, which consisted of only nine questions inquiring about the respondents' perceptions in three areas: (a) management's concern about safety, (b) the effectiveness of safety activities and (c) the existence of physical risks in the workplace. The questions were taken from Dedobbeleer and

Béland (1989). The questionnaire was given to a randomly chosen sample of several respondents in each company, 690 respondents in total.

The 80 companies formed five groups, each representing different implementation strategies. When comparing the implementation groups, the median of each question failed to indicate any differences (Saari *et al.*, 1991). Although the five groups of companies were very different, there seemed to be no difference between them in safety climate.

There were quite clear differences between the groups in safety activities, working conditions, training results and the overall quality of the WHMIS system (Saari *et al.*, 1991). These differences were reflected, for example, in the number of hours spent on various OHS activities and in the health and safety quality of workplaces. Despite the factual differences between the groups, the respondents' median perception about these aspects did not differentiate the groups. On a scale of 0 to 100, the lowest median by implementation strategy was 67 and the highest 73. For example, the number of person hours spent annually on various OHS activities per 100 employees varied from 480 to 1107.

Safety climate was the only instrument that did not show any differences between the groups. As there were quite clear differences between the groups in the objective indicators of safety climate, it can be stated that the instrument of perceived climate failed badly. The shortened questionnaire did not have the same sensitivity as the original one, even if the choice of questions was based on an item analysis. Because the companies were physically in different locations and because, in each company, the questionnaires were given to respondents working in different departments, there could not have been a mutual agreement to answer the questionnaire in a given way. An explanation for the result is that the functional properties of the original instrument were changed too much by reducing the number of questions to a fraction. Dedobbeleer's public comment on this in the Third International Conference on Injury Control and Prevention in Melbourne 1996 was that she did not believe that shortening was the main factor. She instead believed that it was a failure driven by local work culture.

If the malfunctioning of the questionnaire was culturally driven, it demonstrates the hidden risks of transferring measurement tools internationally. Of course, the risk applies primarily to subjective measures. However, there are amazingly big differences also in physical areas. One can wonder how much cultural factors affect such instruments as NIOSH lifting formula, for example. At least socio-demographic factors have a major effect on the applicability of these 'objective' instruments.

In Finland during the late sixties, the author and a Swedish colleague carried out a study on tying reinforced steels in the construction sites (Saari and Wickström, 1978). He brought from Sweden two devices, totally unknown in the Finnish construction industry. With these devices it was possible to tie rods of reinforced steel at floor level in a standing position without needing to bend down. The Finnish workers used a hand tool and wire for tying. When reaching for steel rods on floor, they had to bend down (Sarri and Wickström, 1978). It was suspected that this posture caused excessive musculoskeletal injuries.

Comparative experiments were run with the two new devices and with the traditional method. The results were very clear: both new devices were faster, physiologically lighter and allowed a much better posture. A fairly extensive campaign was arranged to introduce the results and a lot of promotional material was produced for the new devices. The two devices never came into use in Finland.

Almost 30 years later, the Finnish reinforcement workers still use the traditional method. A possible explanation is that the awkward work method gave the workers a better position at the bargaining table. Another explanation is that the Swedish safety culture promotes safety better than the Finnish safety culture (Salminen and Hiltunen, 1995).

When working as an ILO consultant in Bangkok, Thailand, the author visited about 10 large construction sites. In all sites the reinforcement workers were women and tied steel rods with a similar method to that of their Finnish male colleagues. However, there was a major difference in work postures: the Thai workers squatted for tying and did not bend their backs into a bad position, unlike the Finnish workers.

The Swedish devices were unacceptable in Finland, possibly for political reasons. In Thailand, they were not needed. As this example shows, there is an array of factors leading to the rejection of a safety innovation. The totally different work postures used for a similar job in Thailand and in Finland show that the factors go well beyond subjective perceptions and opinions.

13.2.3 Implementation phase

Organizationally good companies implement better

If an intervention has succeeded in all phases, it may still be implemented in very differing ways. There is evidence that companies with good organizational health can implement any new programmes much better than less healthy companies. Marcus (1988) found in a study about new regulations in the nuclear power industry that some companies were able to modify the general requirements of regulations into their local needs while others could not. The number of incidents was clearly lower in the more autonomous companies than in the rule-bound companies.

Similar observations were found in a WHMIS study (Saari *et al.*, 1993). Initially better companies had better implementation results. The best companies reported benefits which went well beyond the scope of WHMIS, for example inventory control, reducing the number of chemicals purchased, etc. The worst companies could hardly develop the new system to comply with the regulations.

The same phenomenon was visible in an intervention study in Finland (Saari, 1986). In this study were companies from four industrial sectors. There were two pairs of companies from each sector. One company in each pair had a lower than average injury rate in its sector and the other company had a higher than average injury rate. One of the pairs served as intervention companies in each sector and the other pair served as comparison companies.

The purpose of the study was to make the work of the local OHS organization more effective. An action package was planned with the safety officers and employees' safety representatives from all intervention companies. Each company took a similar action plan, which they were allowed to modify according to the local needs. The researchers served as resource persons for the participants if help was needed in learning new techniques or in finding background material. Action itself was to be done by the companies.

It was found that companies with a low initial injury rate pushed down their injury rate further but companies that had many accidents initially could hardly implement the action plan. Better companies had no problems in the implementation. Because the companies were from the same sectors of industry and because the pairs were as comparable as possible, it could be seen that the worse companies had all kinds of other problems. Their organizational health was not as good as it was in the better companies.

The actual implementation can vary very much from company to company, depending on the ability of the company to implement a new safety programme. The study from Finland showed this (Saari, 1986) and the same observation has been made in studies comparing different training strategies in OHS (Robins *et al.*, 1990).

A safety programme is implemented in a different way in two countries

In Finland, a workplace improvement process was developed 10 years ago called Tuttava (Saari, 1992). This is a participatory programme which aims at encouraging employees to think about improvements for their jobs and work stations. A team implements Tuttava. The members in the team are workers' representatives, a supervisor and a manager representing higher management. A team implements the process in an area where the number of employees is at least five but not more than 30. Usually, the implementation area is one supervisor's area.

Tuttava has come to have wide use in Finland. More than 1000 companies have used it, almost without exception with good results. When the process was developed several empirical studies were carried out. Accidents went down by 80% in some of these experiments (Saari, 1992). Many companies have seen similar results in their applications.

The main idea in Tuttava is to focus on a theme that is neutral and, for the workers, easy to discuss. The focus is therefore on tools and materials. The team defines what are the good work practices in using tools and in handling materials; usually about 10 practices. In this way they set behavioural targets for the workers. Usually, some of the good work practices cannot be used because of technical, organizational or other obstacles. The team tries to identify these obstacles and takes action to eliminate them. The purpose is to initiate technical improvements along with behavioural changes.

The team also designs a checklist for measuring their performance level. The checklist consists of a large number of items that have only two values: correct or incorrect. The percentage of correct items is a simple indicator for current performance. This number serves as an essential information element to initiate the full adoption of good work practices.

The team makes an observation round in the implementation area on a weekly basis for several weeks. They then call a meeting at which the good work practices, the principles of measurement and the results of baseline measurements are described to everybody working in the area. After the meeting, the results of each observation trip are immediately posted on a large feedback chart in a visible location. This part of Tuttava is quite similar to behavioural safety programmes developed and widely used in North America (Krause *et al.*, 1990; McSween, 1995). Some of these US studies served as the model for using feedback in Tuttava (Komaki *et al.*, 1978; Sulzer-Azaroff, 1978). Behavioural safety programmes and Tuttava differ, however, in some critical aspects. Although both use feedback for initiating the behavioural change, the way feedback is generated is different. In Tuttava, behaviour is not observed directly, only its visible traces at the work station are observed. For example, a typical item on an observation sheet would be 'are cables and hoses coiled if not in use' or 'is cleaning equipment in its designated place if not in use'. Behavioural safety programmes identify and observe the actual behaviours. A Tuttava team observes the marks these behaviours leave at a work station. This makes Tuttava less personal.

Another significant difference is that management, together with experts, implement the behavioural safety programme in US companies. A participatory team implements Tuttava. In the behavioural safety programmes, the behavioural objectives come from accident reports, incident reports or from the management. A behavioural safety programme does not encourage workers to analyse their safety needs. Tuttava teams discuss the best work practices on the basis of the members' daily work experiences. The workers' representatives consult their fellow workers regarding setting the best work practices and the technical improvements required to make these usable. In this phase the workers' role is often more decisive than the managers' role, since workers know the practical needs best.

An experiment with Tuttava is currently underway in the automotive industry in Canada. The test plant is part of a big car company that makes parts for the auto industry. The purpose of this experiment is to test the usability of Tuttava in another country. So far the test has made some cultural differences quite clear (Saari, 1996):

1. In the test plant management–labour relations are significantly different from those in Finland. Traditionally, the managers make the decisions and the workers carry out the decisions. Tuttava calls for joint decision-making. The implementation teams were very quick in identifying problems and even alternative solutions. However, when the teams were supposed to choose between solutions the workers did not participate. They left the decisions to the management. In a Finnish team, the workers would not have avoided taking part in the decisions. Top-down decision-making is traditional in the auto industry and in this corporation. Therefore the workers are used to carrying out decisions the management makes. The different organizational layers have protected and still protect their positions by reprimanding anyone stepping beyond his mandate. Decision-making differentiates the layers.

2. Strict horizontal division of jobs was another significant difference. For example, if a shield slid off its position, leaving a machine operator's legs unprotected from the spilling of the chemical being handled, pulling the shield back to its normal position would be a five-second job and would not require special knowledge or tools. However, the machine operator could not do this himself because it was a maintenance job, even though it was his legs that were at risk. Distinctive job descriptions ensure that everybody knows what his or her job is. If a job is not done, it is easy to identify who is at fault. However, this leads into a slow response in new or abnormal situations as nobody takes care. In the Finnish industry, job descriptions are less specific and more collective by nature. For example, a production worker is expected to do minor maintenance tasks if knowledge and skills are sufficient.

3. The workers did not have either the skills or the social permission to represent their fellow workers. If a representative takes part in a decision that proves to be a less than effective one, fellow workers may blame the representative. This

inhibits the representative's desire to speak out when alternative solutions are analyzed and decisions made. Because it was not usual for the workers in the test plant to represent other workers in various groups, the representatives were uncertain about the consequences of decisions.

These factors made it impossible to implement Tuttava in the normal way. Researchers' roles became much more dominant than in Finland. The researchers had to write down the good work practices, design the observation sheets and the measurement system, and carry out the measurements.

However, Tuttava was implemented and it became successful in changing behaviour and reducing injuries. Good work practices became more frequently used and a large number of technical and other improvements were made. Best of all, the plant reduced its lost-work day injuries by tens of percentage points. The cultural inhibitions meant that the programme had to be implemented in another way. This was possible because a researcher knowledgeable in Tuttava was available all the time.

The starting point of Tuttava was an American safety process (Komaki et al., 1978; Sulzer-Azaroff, 1978; Näsänen and Saari, 1987). Tuttava changed in several aspects, as explained earlier, but some of the original characteristics remained. Even then Tuttava could not be implemented in the same way as in Finland.

13.2.4 Evaluation criteria

It is common in evaluation research to use specific dependent variables to make the effect as noticeable as possible and to help avoid the misinterpretation of results. Therefore criteria more remote in time and contents are often avoided. A follow-up of the WHMIS study shows that the effects of an intervention may become totally different with time (Saari et al., 1993).

The first evaluation round showed that employees knew about the handling of hazardous materials more clearly in companies that had had employees trained by external experts. The companies that used employees trained as trainers did not do as well.

A similar evaluation was carried out three years later in the same companies (Thériault et al., 1995). The picture had turned around. The best knowledge about hazardous materials was displayed in companies using trainers. Employees knew less about hazardous materials in all other companies. The best group, all employees initially trained by experts, had reverted to the same level as other groups. The comparison group, which was not in compliance with training requirements during the first evaluation round, did as well as the companies that had their employees initially trained by external experts.

The results may, of course, depend on the type of safety programme. Safety programmes involving training need time for maturing. On the other hand, any safety programme calls for behavioural changes. Even if the change was purely technical, behavioural changes are necessary either prior to the implementation or afterwards. Therefore almost any safety programme becomes vulnerable to the process of slow changes over time.

13.3 CONCLUSION

Accidents are similar everywhere. A falling person or a finger in a pinch point are injured in the same way despite the location, colour or nationality. Therefore it could be assumed as a first approximation that accident scenarios in different countries vary only because of differences in the type of production, the level of technology, demographic factors, etc. However, there are considerable differences in injury rates in countries and corporations having similar economic and techno- logical structures. The explanation for this is that less successful countries and corporations have not utilized existing safety knowledge efficiently. They could rise to the level of the best companies by doing the same tricks. Examination of safety intervention research in two culturally quite different areas, Scandinavia and Canada, shows that it is very important to understand the cultural standpoint of each country and company before designing safety programmes for them.

Any safety intervention goes through several steps, starting from acceptance and ending with the actual implementation. In each step there are several factors and surrounding conditions that may modify the actual contents and effects of a pro- gramme. These factors extend from political and power conditions to coincidental external factors.

This chapter has primarily dealt with inhibiting factors and failures. This was intentional since the literature typically reports successes only. However, there are cases in which the results are much better than expected and a technique transferred from one cultural area to another works much better than expected (Saari and Näsänen, 1989). It is even possible that a safety programme could work better when transferred to another country or company.

References

DEDOBBELEER, N. and BÉLAND, F. (1989) The interrelationship of attributes of the work setting and workers' safety climate perceptions in construction industry. *Proceedings of the Annual Conference of the Human Factors Association of Canada, Toronto.*

DEDOBBELEER, N. and GERMAN, P. (1987) Safety practices in construction industry. *Journal of Occupational Medicine,* **29**, 863–868.

HALE, A.R. and GLENDON, A.I. (1987) *Individual behaviour in the control of danger,* Amsterdam: Elsevier.

HEINRICH, H.W., PETERSEN, D. and ROOS, N. (1980) *Industrial accident prevention,* New York: McGraw-Hill.

JANSSENS, M., BRETT, J.M. and SMITH, F.J. (1995) Confirmatory cross-cultural research: testing the viability of corporation-wide safety policy. *Academy of Management Journal,* **32**, 364–382.

KAESTNER, N., WARMOTH, E.J. and SYRING, E.M. (1967) Oregon study of advisory letters: the effectiveness of warning letters in driver improvement. *Traffic Safety Research Re- view,* **11**, 67–72.

KIVISTÖ, M. (1995) *Hyvä työpäivä Turun piiritullikamarilla,* Helsinki: Työterveyslaitos ja Valtion Työsuojelurahasto (in Finnish).

KOMAKI, J., BARWICK, K.D. and SCOTT, L.R. (1978) A behavioral approach to occupational safety: pinpointing and reinforcing safe performance in a food manufacturing plant. *Journal of Applied Psychology,* **63**, 434–445.

KRAUSE, T.R., HIDLEY, J.H. and HODSON, S.J. (1990) *The behavior-based safety process,* New York: Van Nostrand Reinhold.

LaBarr, G. (1995) What makes ergonomic teams work? *Occupational Hazards*, **57**, 2.

Laner, S. and Sell, R.G. (1960) An experiment on the effect of specially designed safety posters. *Occupational Psychology*, **34**(3), 153–169.

Liker, J.K., Nagamachi, M. and Lifshitz, Y.R. (1989) A comparative analysis of participatory ergonomics programs in US and Japanese manufacturing plants. *International Journal of Industrial Ergonomics*, **3**, 185–199.

Linstone, H.A. and Mitroff, I.I. (1994) *The challenge of the 21st century: managing technology and ourselves in a shrinking world*, Albany: State University of New York Press.

Marcus, A.A. (1988) Implementing externally induced innovations: a comparison of rule-bound and autonomous approaches. *Academy of Management Journal*, **31**, 235–256.

McSween, T. (1995) *The values-based safety process*, New York: Van Nostrand Reinhold.

Näsänen, M. and Saari, J. (1987) The effects of positive feedback on housekeeping and accidents at a shipyard. *Journal of Occupational Accidents*, **8**, 237–250.

Rafnsson, V. (1993) Risk of fatal accidents occurring other than at sea among Icelandic seamen. *British Medical Journal*, **306**, 1379–1381.

Robins, T.G., Hugentobler, M.K., Kaminski, M. and Klitzman, S. (1990) Implementation of the federal hazard communication standard: does training work? *Journal of Occupational Medicine*, **32**, 1133–1140.

Saarela, K.L., Saari, J. and Aaltonen, M. (1989) The effects of an informational safety campaign in the shipbuilding industry. *Journal of Occupational Accidents*, **10**, 255–266.

Saari, J. (1986) Tehoa tapaturmantorjuntaan. *Työterveyslaitoksen tutkimuksia*, **4**, 261–430 (in Finnish).

(1992) Scientific housekeeping studies. In *Profits are in order*, ed. F.E. Bird, Jr, pp. 27–42, Atlanta: International Loss Control Institute.

(1995) Risk assessment and risk evaluation and the training of OHS professionals. *Safety Science*, **20**, 183–189.

(1996) Use of a safety program in two different organizational cultures. Presentation at CybErg 96, http://www.curtin.edu.au/conference/cyberg/.

Saari, J. and Näsänen, M. (1989) The effect of positive feedback on industrial housekeeping and accidents: a long-term study at a shipyard. *International Journal of Industrial Ergonomics*, **4**, 201–211.

Saari, J. and Wickström, G. (1978) Load on back in concrete reinforcement work. *Scandinavian Journal of Work, Environment and Health*, **4**, 13–19.

Saari, J., Altonen, M., Kopperi, M., Lehtonen, K., Simpanen, V. and Seppänen, S. (1983) Model for the investigations of occupational accidents, presentation at the specialist day: Analysis of the risk of accidents at work applications, Ottawa, May, pp. 117–132, Geneva: ISSA and ILO.

Saari, J., Bédard, S., Dufort, V., Hryniewiecki, J. and Thériault, G. (1991) *Évaluation de límplantation de mesures de sécurité initiés de léxterieur – le cas du simdut*, Montréal: École de santé au travail, McGill University.

(1993) How companies respond to new safety regulations. *International Labour Review*, **132**, 65–74.

(1994) Successful training strategies to implement a workplace hazardous materials information system. *Journal of Occupational Medicine*, **36**, 569–574.

Salminen, S. and Hiltunen, E. (1995) Cultural differences in occupational accidents: Part 2. A case study of Finnish and Swedish speaking workers in Finland. Presentation at the 44th Nordic Meeting on Work Environment, Naantali, August.

Seppälä, A., Saarela, K.L., Näsänen, M., Aaltonen, M. and Saari, J. (1987) Improving safety performance of industryi. In *Trends in ergonomics/human factors IV*, ed. S.S. Asfour, Amsterdam: Elsevier.

Sheehy, N.P. and Chapman, A.J. (1987) Industrial accidents. In *Industrial Review of Industrial and Organizational Psychology*, eds G.L. Cooper and I.T. Robertson, pp. 201–227, Chichester: Wiley.

SUCHMAN, E.A. (1965) Cultural and social factors in accident occurrence and control. *Journal of Occupational Medicine*, **7**, 487–492.

SULZER-AZAROFF, B. (1978) Behavioral ecology and accident prevention. *Journal of Organizational Behavior Management*, **2**, 11–44.

THÉRIAULT, G., FERRON, M., DUFORT, V., BÉDARD, S. and SAARI, J. (1995) *Performance à moyen terme de divers modes d'implantation du SIMDUT*, Montréal: Département de santé au travail, Faculté de médecine, Université McGill.

TUOMINEN, R. and SAARI, J. (1982) A model for analysis of accidents and its application. *Journal of Occupational Accidents*, **4**, 263–273.

CHAPTER FOURTEEN

Before it is too late: evaluating the effectiveness of interventions

ANDREA SHAW AND VERNA BLEWETT

14.1 INTRODUCTION

When we intervene to prevent occupational injury, we are taking action in work-places. For these interventions to be durable within the competing imperatives of the workplace, their effectiveness must be demonstrable. This requires evaluation: 'the collection and analysis of information in order to facilitate informed decision making' (Lambert and Owen, 1992). Traditionally, OHS researchers have used accident data as the primary, often even the sole, evaluation criterion. In this chapter, we argue that reliance on such a single outcome measure cannot provide adequate evaluation information nor indeed support successful interventions. An injury-prevention intervention in a meatworks illustrates this argument. In this meatworks, the OHS committee had determined to change working arrangements in the work area where sheep are killed and processed in order to address manual-handling risks associated with pelting (removing the skin of the beast). In intro-ducing these changes, there were a number of industrial sensitivities. The company was also concerned that the changes did not cause damage to the skins, which are sold as, of course, sheepskins. The committee therefore decided to implement the changes on a trial basis. They then faced the problem of how to decide whether the trial had worked. The committee developed a series of indicators:

- *Number of injuries*: collected according to the type of injury (e.g. cuts, sprains and strains) and whether it was a new injury or a recurrence of an existing injury. Rather than calculate frequency rate, which the committee found difficult to use, the number of injuries was matched against the number of people working in the area. This information was collected by the supervisors and the OHS officer.
- *Take-off of the skins*: the percentage of the skins of the required quality for sale. This information was collected by the skin coordinator.
- *How people felt about the changes*: according to a scale from 'heaps better' to 'heaps worse'. This information was collected by the union delegate and the supervisors.

The committee agreed to trial the changed working arrangements for three months. After the first six weeks, the committee reviewed the changes and performance against the indicators in order to modify the changes. After six weeks, the committee found:

- A clear improvement (i.e. a reduction) in the number of injuries. Over the most recent three weeks, there had been no injuries of any kind.

- Skin take-off had deteriorated briefly. However, the skin coordinator and the team explained this as the result of 'dry sheep', a seasonal problem which they had experienced at this time of the year before. In fact, the skin coordinator expressed the belief that skin take-off had not deteriorated quite so much this year.

- Everyone agreed that the work had got 'heaps better', despite the initial concerns. However, the supervisors had some practical problems with the changed working arrangements which the committee were able to address.

At the end of the trial, the trends that were apparent at the six-week review had continued:

- While there had continued to be some injuries, the rate had clearly dropped.

- Skin take-off had not deteriorated any further and had, in fact, improved as the quality of the sheep had improved.

- The union delegate and the supervisors reported that the workers in the area could not imagine doing the work any other way.

This example illustrates a number of the consequences for evaluation that intervention in the workplace entails:

- Workplaces are complex sociotechnical systems. This means that interventions usually have to meet a number of potentially competing imperatives. In the example of the meatworks, the imperatives included maintaining the quality and speed of production, respecting existing industrial relations agreements about working arrangements as well as controlling OHS risks in the area.

- The cause and effect relationship between interventions and outcomes is often not immediately apparent. For example, manual-handling interventions can take many years to have full effect and occupational diseases can have latent periods in excess of 20 years. Despite the immediate reduction in injury rates in the mutton area in the meatworks, it is unlikely that the full effect of the intervention will be expressed for some time, particularly all benefits in relation to manual handling.

- In any workplace, interventions must be able to be evaluated as you go, to fine-tune interventions, to identify and address confounding factors and to build preparedness to at least try a new method of working. In the above example, the supervisors' practical problems with implementing the change had to be solved for the otherwise successful intervention to continue.

Traditional approaches to evaluating the effectiveness of interventions which rely on accident data cannot take all of this into account. Accident data alone will not allow us to evaluate the intervention's effects (for good or ill) on other imperatives such as quality and productivity. Nor will they tell us if concurrent workplace changes to address, say, productivity are confounding the effects of our occupational injury intervention. A new approach to evaluation is needed that recognizes all of the imperatives that injury-prevention interventions face in workplaces. To develop such a new approach, we need to look closely at how injuries occur.

14.2 THE ASET PROCESS

Occupational injuries are the result of a series of events that occur in organizations. They result from exposures to hazards existing in the workplace environment. These in turn arise from the systems used to manage a workplace, which are determined by the culture of the enterprise. A further example illustrates this. In a medium-sized manufacturing plant, the forklift driver was delivering work in process from one part of the plant to another using the recently purchased forklift truck. The driver stopped to talk to one of the workers and during the conversation product fell from the forks hitting the pedestrian worker and injuring her shoulder. The plant had clear procedures for first aid and accident investigation, which went into full swing immediately. The initial reaction was to find fault with the driver as she was new to the equipment, but later investigation revealed a fault with the lifting mechanism of the forklift. The response was to repair the forklift immediately and checks were also made of similar machines to ensure that the same fault was not widespread. Such action was appropriate and would normally be applauded as being the right response to an industrial accident.

Unfortunately, the accident investigation did not reveal the whole story and repairs to the forklift did not repair the causative agents that were systemic in the plant. A look behind the scenes reveals a whole new perspective on this event. The company competed with other suppliers solely on the basis of the cost of its product and it had a philosophy, or a set of values, which supported this business strategy. The company valued cheap price over quality, believed that 'cheaper' also meant 'better' and rewarded employees for cutting corners. These underlying cultural features in the company led to the development of management systems that supported this philosophy. Consider, for example, its procurement system: in order to maintain its position as a low-cost supplier of its product, the company fought to keep its inputs to production as low as possible. When it was clear that a new forklift was needed to deal with expansion in the business, the purchasing officer naturally took advantage of the offer of a second-hand unit from the local dealer. The machine was cheap because it carried no warranty and was offered on the basis of *caveat emptor*. Therefore the system that was ostensibly designed to allow the company to compete ultimately resulted in the worker's injury.

The accident investigation that took place was an effective examination of events leading up to the accident but, as usual, did not include an examination of the systems impacting on the event or the cultural elements that may contribute to events. Without this examination of upstream events, the investigation, as we have seen, is incomplete. No one asked the questions:

- What failed in our purchasing system that allowed us to purchase a faulty forklift?
- What is there in our culture that allows us to adopt systems that lead to our people being injured?

This link between culture, systems, exposures and injury is a little like a river: what happens upstream has consequences for the river downstream; heavy rain at the head of the river gives rise to flooding further down. If the flooding is to be controlled, action needs to be taken upstream rather than downstream where the flood waters have had the chance to spread. Like the river, health and safety issues also need to be resolved upstream rather than downsream. That is, effort needs to be put into examining the atmosphere or culture of the organization and the systems that arise from it. Upstream factors are predictive of downstream events and can help control them.

Figure 14.1, based on the work of Krause and Finley (1993), represents the relationship between the culture or atmosphere in an enterprise, the systems it adopts, the various hazards to which people are exposed and the incidents that result in ill health or injury, that is the targets that most enterprises measure as the bottom line of their OHS management practices. While Figure 14.1 represents the flow as one way, clearly feedback loops exist. For example, changes to the systems used to investigate accidents can support change to a belief that accidents are caused by careless workers.

This model provides a framework for examining and identifying needed improvements in key processes relating to OHS. It supports identification of key features in the upstream, focusing attention onto the precursors of occupational injury.

For example, in one enterprise, quite sophisticated OHS management systems in most areas did not lead to substantial improvement in exposures due to behaviour nor to reported lost-time injuries. By investigating features of the enterprise's OHS culture through focus group interviews with employees, some of the reasons for this apparent paradox became clear. Despite public commitment to OHS and the implementation of considerable improvements to OHS infrastructure in recent years, most employees did not believe that management took OHS seriously or that they 'really cared' about the workforce. Vigorous action as a result of lost-time injuries was cited as evidence that 'someone has to get hurt for anything to change around here.' Lack of trust between management and workforce was a prominent feature; management were suspicious of many workers' compensation claims and the workforce did not believe that management commitment to OHS was sincere. OHS was seen as an area for competition between management and the workforce and was the source of some conflict.

Clearly, in order for this enterprise to reap the benefits of the improvements they had already made and intended to make in the future to OHS management, they needed to address cultural issues. Through identifying the issues, some strategies became immediately apparent. For example, the organizational reaction to lost-

Atmosphere →	Systems →	Exposure →	Target
eg vision, values	eg training,	eg state of	eg incidents,
common goals	purchasing,	equipment,	near misses
beliefs	hazard control	behaviour	
	procedures		

Figure 14.1   The ASET process.

time injuries was dampened and more senior management involvement in regular hazard identification activities began. The focus group interviews themselves identified a number of strategies for improvement, such as greater management visibility in day-to-day operations, particularly relating to OHS.

In another case, detailed examination of the exposures of a workplace revealed the ways in which atmosphere had formed systems that had in turn resulted in exposure to significant risks. Management and workforce shared a strong belief that this workplace was necessarily very dangerous: 'we don't work in a chocolate factory'. This belief was cited as justification for not developing and implementing effective systems for dealing with many of the hazards of the work process. Risk control measures were only put in place as a reaction to incidents, which were treated as evidence that the danger had become too great.

By looking more deeply than just the exposures, examining the systems and cultural features that had lead to the exposures, the investigation was able to provide recommendations that acted at a number of levels. As well as controlling risks, strategies that increased operator participation and control of the working environment were recommended. The need to demonstrate that risks can and should be controlled was addressed by recommendations for straightforward and immediate risk control measures. Finally, the implementation of a range of systems necessary for risk control was also recommended.

As both these examples demonstrate, using the ASET model provides useful insight into enterprises, illuminating features of OHS management on several levels. It leads to the development of injury-prevention interventions that act on a range of issues at all levels, supporting greater and more durable benefits from the intervention. Evaluating these sorts of interventions requires a more sophisticated approach than just measuring lost-time injury frequency rates. Furthermore, given such cultural environments, evaluating traditional injury-prevention interventions requires an examination of their effect on the atmosphere and systems to ensure that future problems have not been created. Such evaluation will necessarily include subjective, as well as objective, information.

14.3 ACCIDENT DATA ARE POOR INDICATORS OF IMPROVEMENTS

One of the common justifications given for using accident data as the sole measure of OHS performance is that accident data are objective, as opposed to qualitative measures, which are subjective. However, if the key purpose of evaluation is to improve the effectiveness of interventions, then the objective–subjective distinction is not useful. The most important feature needed of evaluation measures is reliability. If the measure is reliable, then the same information will yield the same measure, regardless of who is actually making the measurement.

In fact, accident data, such as lost-time injury frequency rates (LTIFRs) are not reliable without supplementary data. Amis and Booth (1992) point out that, as with any set of statistical measures, accident data are subject to random variation. Unless a variation is sufficiently large or a rate change is sustained over a sufficiently long period, a rise or fall in, say, LTIFR may be nothing more than a variation within the standard deviation and nothing at all to do with the intervention under investigation. In workplaces, the change may even be due to the so-called 'Hawthorne effect' (Bailey *et al.*, 1991). Accident data alone will not demonstrate any link between a drop in LTIFR within standard deviation and a particular injury-prevention intervention.

Reporting criteria can also make accident data less than reliable. Haines and Kian (1991) suggest that a reliance on accident data can even inhibit full accident reporting. Anecdotal evidence from our own work supports this view. At the very least, given that any evaluation will have to rely on information from workplaces, if there are different approaches to reporting and recording of information in different work areas or plants, the accident data may not be comparable. For example, attitudes and conventions can lead to under- or over-reporting. Different approaches to rehabilitation can affect LTIFRs in particular. One company's lost-time injury may be another company's modified duties case.

Because the workplace is a complex system, there are usually problems directly linking an intervention with immediate outcomes. As described above, many injury-prevention interventions take a considerable period of time, up to years, to yield outcomes. Using accident data as the sole measure of effectiveness raises problems because the information is needed within a shorter period of time. For example, interventions to deal with manual-handling injuries may not show their full effect for some 10–15 years. However, evaluations will be needed within that time period to allow modification and improvement of the intervention strategies. Furthermore, relying on accident data as the sole evaluation measure will not provide reliable information because any manual-handling injuries may in fact be due to the situation *prior* to the intervention.

Finally, accident data do not allow links to be drawn between injury-prevention programmes and other changes in the workplace. For example, the injury-prevention programme may be addressing manual-handling risks effectively, but speeding up the production line as a result of higher productivity targets may more than offset any gains. Using accident data as the sole evaluation measure may not necessarily identify this confounding effect (see Kletz (1993) Shaw and Blewett (1995) Worksafe Australia (1994) for further discussions of problems with accident data).

14.4 WHAT'S NEEDED?

Accident data only provide information about the last step in the ASET process, i.e. targets. For reliable information which allows evaluation and improvement of injury prevention interventions, measures of performance at the earlier stages of the injury causation process are necessary. These allow the links between the intervention and the outcome (reduction in injury) to be demonstrated. They also allow more timely information collection and can help draw out the consequences of other organizational changes.

Measuring the performance of atmosphere and systems (or, at times, exposures) requires the development of performance indicators, often called positive performance indicators in the context of OHS. A range of performance indicators is required to give a holistic view of an intervention's effect in an enterprise. This will necessarily include a combination of measures that have upstream relevance as well as target or outcome relevance. These can provide quantitative information, such as the number of interventions implemented or the number of training courses successfully completed, or they can be qualitative, e.g. assessment of the effectiveness of a training course.

Traditionally, we have come to rely on quantitative data as the predominant form of performance measure; by their very nature numbers seem to carry the weight of validity and reliability. However, qualitative information is also an important feature. Indicators such as satisfaction with an injury-prevention programme can provide useful information to help refine and improve strategies. For example, if 80% of employees report that they find a new manual-handling procedure difficult to implement, an evaluator can reliably conclude that the intervention is not working as it should, even though each employee has formed their opinion for subjective reasons. Qualitative data are also a rich source of information in their own right. For example, Chapter 2 relates how narrative fields on injury data provide the means to focus prevention activity in a way that the numbers alone could not do.

A range of measures is needed, each like a piece of a jigsaw puzzle, which, when put together, builds a picture of the whole enterprise. The main issue is reliability; that is, getting the same result from measures by different people. Reliability is dependent on the measurement tools developed, particularly for qualitative data. The use of clear surveys, unambiguous interview questions and consistent data collection protocols is critical. Finding reliable measures also depends to some extent on the process used to select the measures. We advocate a consultative approach where users and those who will collect the data have a say in developing the measures. Such an approach ensures that all key features of the intervention being evaluated are addressed by the performance indicators and that the approach is congruent with organizational culture. On a practical level it also makes the use of the performance indicators more straightforward.

14.5 HOW DOES IT WORK?

Developing performance indicators for occupational injury-prevention interventions must be based on an explicit understanding of the range of goals an intervention must achieve. This will not necessarily be straightforward; it relies on a detailed understanding of the workplace environment, reinforcing the need for a participative approach. Once these goals have been identified, indicators which show how effectively the intervention is meeting these imperatives can be developed. However, as the name suggests, performance indicators will only *indicate* how effectively a process is operating. They are not necessarily perfect measures. Performance indicators will not always tell the complete story, but they will indicate where to start to improve performance.

To develop such performance indicators, a group process that involves users is most effective. Tools such as brainstorming and consensus decision-making are used to answer the question 'How do we know how well this intervention is operating?' From a long list of possible indicators, the group chooses those which best focus on the *precursors* of good performance, the lead indicators, and those which tell how effective the intervention is rather than merely how much work it has required (performance as opposed to 'busyness' indicators). The final performance indicators must be sufficient to allow necessary improvements in the intervention to be identified, but not so many that the evaluation becomes impractical.

The example with which this chapter began provides a simple example of the types of indicators that are useful in the evaluation of interventions. As the example describes, using the range of performance criteria allowed the intervention to be fine-tuned to increase its effectiveness. It also ensured that the intervention did not

Table 14.1 Examples of positive performance indicators

Type of intervention	Performance indicators
Improved hazard identification processes	Percentage of identified problems rectified within agreed timeframe Percentage of inspections completed on time Ratio of repeated problems identified to total number identified
Training	Participant assessment of the benefit of the training Percentage of participants who pass relevant competency assessments Percentage of designated employees attending training programme
Employee induction	Turnover of new employees (within six months) Percentage passing competency assessment for new employees Supervisor assessment of the benefits of the new induction process
Improved consultation	Percentage of committee actions completed on time Ratio of actual to expected management attendance at committee meetings Percentage of workforce participating in OHS activities Assessment by workforce of the effectiveness of consultation

adversely affect productivity, quality and industrial relations. If adverse effects had occurred, the intervention would have been unlikely to continue, regardless of its effect on accident rates.

In another enterprise, management commitment to OHS was identified as a key constraint on OHS performance. The enterprise determined to implement strategies for reinforcing and demonstrating this commitment. They were able to show the success of their strategies by measuring:

- the number of field visits made by senior managers; and
- the employees' perception of the commitment of senior managers to OHS as determined by an attitude survey.

Movements in these indicators allowed tracking of progress in implementation of the strategies: How do perceptions vary across the workforce? Are all senior managers making the same number of site visits? Are variations in these indicators related? They also allow identification of further possible interventions, such as greater concentration in particular work areas.

Table 14.1 provides further examples of performance indicators that have been developed using the participative process described above to be used to supplement accident data in evaluating prevention interventions.

14.6 CONCLUSION

The primary focus of the evaluation of injury-prevention strategies must be the improvement of the effectiveness of interventions. This focus requires that evaluation measures are reliable and can be used to revise strategies or identify the need for new ones. By supplementing accident data with indicators of the performance of upstream systems and activities, an evaluation system that allows improvement can be established. For this system to be of most value in workplaces, it must be developed in a highly consultative process. This ensures that performance indicators are understood and can be acted on at all levels in the enterprise.

Such an evaluation system can support improved occupational injury-intervention programmes. By providing meaningful and practical performance measures, commitment to and support for interventions can increase. Such a system also allows occupational injury-prevention strategies to be linked into other business improvement strategies, which are often perceived to be more central to the enterprise's activities. For example, in our experience, support for an intervention strategy has been reinforced when positive effects on workplace relationships were evident from an evaluation, regardless of immediate effects (if any) on accident data.

In summary, an evaluation system which includes measures of the precursors of accidents allows much more effective targeting of prevention activities, before it is too late.

References

AMIS, R.H. and BOOTH, R.T. (1992) Monitoring health and safety management. *The Safety and Health Practitioner*, Feb., 43–46.

BAILEY, J.E., SCHERMERHORN, J.R., HUNT, J.G. and OSBORN, R.N. (1991) *Managing organisational behaviour*, Milton, Queensland: Wiley.

HAINES, M.R. and KIAN, D.V.S. (1991) Assessing safety performance after the era of the LTI. In *Proceedings of the First International Conference on Health, Safety and Environment in Oil and Gas Exploration and Production*, pp. 235–242, Richardson, Texas: Society of Petroleum Engineers.

KLETZ, T. (1993) Accident data – the need for a new look at the sort of data that are collected and analysed. *Safety Science*, **16**, 407–415.

KRAUSE, T.R. and FINLEY, R.M. (1993) Safety and continuous improvement – two sides of the same coin. *The Safety and Health Practitioner*, Sep., 19–22.

LAMBERT, F. and OWEN, J. (1992) *A guide to program evaluation*, Canberra: Department of Employment, Education and Training.

SHAW, A. and BLEWETT, V. (1995) Measuring performance in OHS: using positive performance indicators. *Journal of Occupational Health and Safety, Australia and New Zealand*, **11**(3), 353–358.

STOUT, N. (1996) Analysis of narrative text fields in occupational injury data, this volume.

WORKSAFE AUSTRALIA (1994) *Positive performance indicators. Part 1 – issues, Part 2 – practical approaches*, Canberra: AGPS.

CHAPTER FIFTEEN

A three-dimensional model relating intervention and cooperation to injury prevention: background, description and application

EWA MENCKEL

15.1 INTRODUCTION

Broadly speaking, there have been two trends in recent injury-prevention research in the occupational arena: one based on the utilization of the increasing range and power of facilities available for data storage and processing, the other based on the realization of the sometimes highly specific nature, to workplace, equipment or task, of occupational-injury problems. The former has acted as a stimulus to what might be called 'analytic' models *of* the injury phenomenon and their statistical testing, the latter to models *for* specific interventions in the workplace, usually on the basis of some form of collaboration between different sets of stakeholders (parties with any kind of material interest). This chapter, although aspects of the status of intervention research per se will not be examined (e.g. see Goldenhar and Sculte, 1994; Westlander, 1993), is largely concerned with the latter kind of model.

It is the author's contention that, as data-processing and analytic models have become ever more sophisticated, increasing difficulties have arisen in applying them to specific kinds of activities and in individual workplaces. Some of these models have required expert knowledge and have not been universally applicable. Recognition of these defects has meant that interest in workplace-specific models and methods has increased, at least in Sweden (Menckel and Kullinger, 1996). These are being developed on the basis of specific environments, people and resources. Although there has been no dramatic change in perspectives on how injuries arise and how they can be prevented, recent occupational-injury research has acquired closer ties with individual workplaces.

An emphasis now seems to be being placed on developing a more coordinated mode of intervening to prevent injuries. In the first instance, this involves direct collaboration between occupational researchers and practitioners, in which researchers act as a support for practitioners in both work to promote change and development and in evaluation of activities undertaken.

Incorporating the notions of intervention, cooperation and prevention (and their elements) into a coherent model has been the author's concern in much of her recent work (Menckel, 1990, 1993). The three-dimensional outline model presented here can be regarded as an attempt to pull some of the research strands together. Presentation of the model is preceded by some background remarks, and followed by an outline account of one of its practical applications.

15.2 BACKGROUND

As our knowledge of occupational injuries has increased, so too has the depth and scope of occupational-injury research. Injuries are no longer considered as isolated happenings, but as the culmination of a process comprising several stages or sub-events. Understanding the diversity in this process provides the foundation for analysis of connections and interacting factors.

As research has progressed in the occupational-injury arena an ever more complicated phenomenon has emerged as its object of study. It has became apparent that there is a need to identify common features, similarities and patterns in injury causes, event sequences and consequences. It is evident that both situational and structural factors need to be analyzed in injury genesis. Systemic risk management (comprising hazard identification, risk analysis and the taking of preventive counter-measures) can then be regarded as an integrated part of the work situation and the operations of an employing organization.

In this context, work-organizational factors have been increasingly focused on. Such factors have been studied at three levels: within the operations of a producing organization as a whole, during a production phase or in a production unit and in the

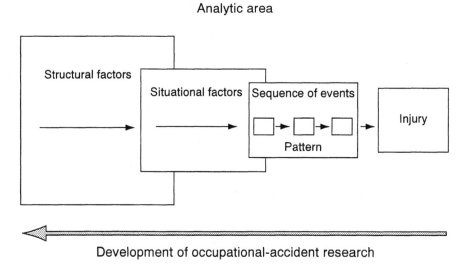

Figure 15.1 Development of occupational-injury research. Source: Laflamme and Menckel, reproduced in Menckel and Kullinger, 1996.

workplace, i.e. where the injury was incurred. The development of the systems perspective on injuries, i.e. one that incorporates both causal mechanisms and context, is illustrated in Figure 15.1.

In analyzing an occupational injury that has already been sustained, it is still customary to proceed in backwards steps from the injury itself, via more proximate events and phases, to the factors determining the immediate circumstances under which an injury was incurred. Accordingly, the usefulness of a model *of* occupational injuries may be very much dependent on an effective model *for* injury prevention in the workplace.

15.3 A THREE-DIMENSIONAL MODEL FOR INJURY PREVENTION IN THE WORKPLACE

The model for injury prevention presented here is shown in Figure 15.2. It relates intervention and any accompanying inter-stakeholder cooperation to prevention.

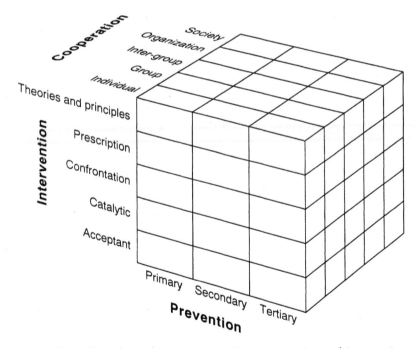

Figure 15.2 The relation between prevention, cooperation and intervention.

Cooperation and intervention are cornerstones in work for injury prevention, and it is in the individual workplace that researchers and practitioners encounter each other in preventive efforts. However, the concepts of prevention, cooperation and intervention can have different meanings according to context and to who is actually involved.

15.3.1 Prevention

The concept of prevention in a health context is both old and new: old in that people throughout the ages have attempted to protect themselves against ill-health; new in that its meaning has tended increasingly to change. The emphasis has shifted from 'protecting' against a known hazard to 'taking action in advance', and the concept has also come to be applied in new and wider contexts. There has, for example, been a shift in emphasis from avoiding specific states of ill-health to the promotion of health from a community perspective.

When the term 'prevention' is employed in everyday speech, it can be hard to distinguish between usages where it refers to goals, such as those of a prevention programme, and other usages where it refers to methods for achieving such goals. In everyday Swedish, the terms equivalent to 'prevention' and 'preventive work' are usually employed to refer to efforts made in advance. By contrast, in the medical research tradition the prevention arena also encompasses actions to restore health when it is impaired, and to minimize suffering and distress, i.e. measures (such as treatment or rehabilitation) that are taken after an injury has been sustained.

In terms of time, a distinction is usually made between three levels of prevention: *primary prevention*, i.e. reduction of the risk of ill-health, *secondary prevention*, i.e. reduction of the prevalence of ill-health and *tertiary prevention*, i.e. reduction of the consequences of ill-health.

The concept of prevention has been extended over the years. From having been directed at a *single* disease, factor or individual it has come to refer to more complex states of disease and multifactorial sets of relationships; it encompasses many or all people in society and also their total environment. Its primary-preventive component, that of intervening before an injury arises, has been increasingly emphasized, so much so that in recent conceptualizations primary prevention is regarded as virtually synonymous with prevention itself. From having principally been documented within one discipline, namely medicine, preventive work has come to require a multidisciplinary approach.

The traditional classification of preventive activities worked well as long as it was applied to hindering the occurrence of specific diseases with known causal factors, e.g. malaria. It becomes considerably more problematic when preventive work addresses new areas, such as the occupational environment, where there is a different time dimension and greater complexity, and where account needs to be taken of a larger number of interacting causal factors.

In recent decades there has been an increasing concentration of interest on primary prevention. Two types of primary-preventive measures are relevant to local preventive work:

1. *proactive primary-preventive measures*: designed to hinder the appearance of a hazard in the environment;

2. *reactive primary-preventive measures*: designed to strengthen the individual so that a risk can be coped with more effectively.

Proactive measures focus on improvements to the physical environment, machines, tools, working methods and work organization, while reactive efforts are aimed at the individual. Examples of the latter include the provision of information to employees on occupational-injury risks and training in safe behaviour.

A number of models of prevention have been developed and tested over the years. There are at least 11 commonly employed concepts of prevention in the research literature, and comparisons between them and their related models have recently been made (Andersson and Menckel, 1995). Five of these can be characterized as being general by nature, while six are clearly specifically related to unintentional injuries. The most striking feature that the different concepts have in common is that they all incorporate a time dimension. Other similarities concern the direction of preventive measures (bottom-up or top-down), the factor at which they are aimed (e.g. host, agent or environment) and at what level they are located (individual, group, organization or society). Similarities of these latter kinds are to be found primarily between the models oriented towards injuries rather than diseases. The various dimensions, one of which clearly involves differentiating between kinds of interventions, are summarized in Figure 15.3.

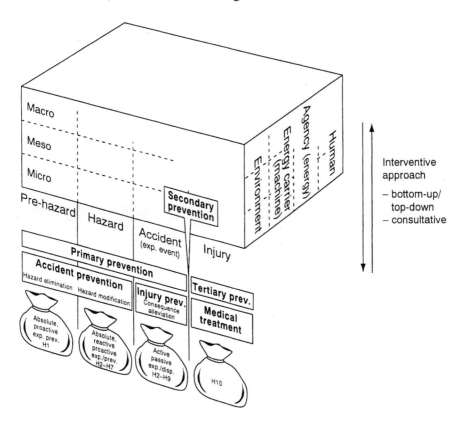

Figure 15.3 A composite model for injury prevention. The 'sacks' contain concepts related on other dimensions than time. HI-HIO refer to Haddon's ten strategies. Source: Andersson and Menckel, 1995.

15.3.2 Cooperation

Cooperation, in particular between researchers and practitioners, is a key element in long-term and durable preventive work. Such cooperation can take different forms of expression: everything from researchers studying processes of change, through researchers functioning as a source of support for organizational or workplace development, to active collaboration between researcher and practitioner in the course of preventive activities. One component of long-term preventive efforts involves contributing to the development of potential resources for safety work, resources that can be adapted to meet other needs and address other problems better.

Cooperation with a preventive purpose can take place both before or after an injury has been incurred. The work can be aimed at the work environment, the work organization and/or the individual. It can be conducted at national, regional or local level, and oriented towards an organization, group or particular individual. Collaboration between researchers and between researchers and practitioners in the workplace is especially interesting, because, in the latter case, it is in the workplace that practitioners can contribute personal knowledge of their own work environment and then apply the methods that have been developed.

Cooperation, whatever its form of expression, involves interaction between different people. Some examples of its forms can be utilized to discuss how it works in practice:

- *information work*, i.e. activities undertaken to disseminate information between different stakeholders, e.g. on injuries that have arisen, the magnitude of risks in different areas, and/or measures taken and their results;

- *parallel work*, i.e. activities where two or more stakeholders work in parallel but where the active integration of knowledge is not necessarily required, e.g. in gathering information on workplace hazards, in 'traditional' safety inspection rounds, or in compilations of knowledge on the basis of the particular area of competence of each individual concerned;

- *collaborative work*, i.e. activities where the knowledge of more than one stakeholder is actively welded together and analyzed communally, e.g. in joint risk surveys and the taking of follow-up measures, in method development or research work, or in compilations of existing knowledge;

- *teamwork*, i.e. activities where the knowledge of more than one researcher and/or researcher and practitioner is required, and actively integrated in relation to a specific pre-set goal or on the basis of a common theory, e.g. in cross-disciplinary programmes specifically designed to reduce the number of occupational injuries.

Of these different forms of cooperation, teamwork is regarded as the most integrated and coherent in relation to an overall target. However, it is also the most complex and appears to be the most difficult to manage in practice. It is easier to find examples of information-oriented cooperation, which is also the case for what we have called parallel work and collaborative work. In practice, however, although the boundaries are fluid, the differences formulated above can still be used as a basis to discuss cooperation and the forms it can take.

15.3.3 Intervention

The third cornerstone, intervention, can be regarded as requiring a person actively to step in to prevent an accident from occurring. Blake and Mouton (1989) define intervention as 'activities/actions' by means of which a 'consultant' can assist a 'client' in solving his/her problems. The term 'client' encompasses both an individual and an entire organization. Note that the emphasis is on 'to modify or hinder' a course of events rather than 'to interfere' (which is the widely understood meaning in an international political context).

For Blake and Mouton, intervention can be directed solely at human behaviour, either that of individuals or of members of a group or of people in a wider organization. They regard behaviour as cyclical by nature, by which they mean that the behaviour recurs at certain intervals, either with a great degree of regularity or more infrequently and unevenly. Intervention is employed to break a cyclical pattern of negative behaviour – a vicious circle. Even though Blake and Mouton speak of intervention principally in terms of activities that are directed at the human being or human behaviour, parallels of relevance in the area of the work environment and to different forms of preventive work can be drawn.

Injuries are recurring phenomena, with varying intervals between them; countermeasures are taken and then have to be repeated. Personnel who are aware of hazards in the occupational environment leave their job and are replaced by people with only limited knowledge. Problems related to an unsatisfactory work environment are concealed for a while but then re-appear in new guise. Work conditions change and new technology is introduced. In this context, an interventive attitude, with its emphasis on an active approach, can help to break what appears to be regular course of events and pave the way for the achievement of more permanent solutions to the problems that arise.

Blake and Mouton offer a perspective on and classification of interventions and different forms of interaction that can be of benefit in characterizing and discussing the preventive efforts of both researchers and practitioners. The authors distinguish between what they call five kinds of interventions:

1. *theories and principles*, e.g. through the provision of training and advice, people can be provided with certain perspectives which enable them to handle their problems themselves;

2. *prescription*, e.g. through the giving of instructions or recommendations, people can be told what should be done in certain situations;

3. *confrontation*, e.g. through demonstrating/displaying different interpretations or contrasting options, people can obtain a better understanding of a problem and their own role in relation to it, so that they themselves may find a solution;

4. *catalytic*, e.g. through supportive influence, people obtain a more extensive knowledge of a problem so that they come to see both its causes and possible solutions;

5. *acceptant*, e.g. through the provision of assistance and support, people are provided with a secure base from which they dare to express their own views on a problem and its solution.

In a practical context, the different kinds of interventions are generally used in their pure form only to a limited extent. To distinguish between them theoretically can nevertheless facilitate and provide an aid to discussion on how a particular problem should be handled. In practical work, it can also be the case that the use of a particular kind of intervention may be appropriate at one stage in the development of a problem while another kind of intervention should be applied at a later stage.

The different kinds of interventions can also be applied at different levels, i.e. the intervention may be directed at a particular individual, at a group of individuals, at the relation between groups of individuals, at an organization or at a wider system, e.g. at community level. They can also serve as bases for how preventive work can be set up in practice. The different kinds of interventions offer varying scope for active work and therefore for different kinds of preventive efforts on the part of researchers and practitioners. Acceptant intervention at an individual level, for example, refers primarily to secondary and tertiary preventive activities, while theory and principles at an individual level give greater scope for the adoption of a primary-preventive approach.

15.4 AN EXAMPLE: MUSCULOSKELETAL INJURIES IN HEALTH CARE

Work-related injuries to the muscles, joints and skeleton are a major problem throughout the world. They are common causes of enduring pain, functional disability, sick leave and premature retirement. Women in health-care occupations run a particularly high risk (Hagberg *et al.*, 1995; Engkvist *et al.*, 1992).

The scarcity of intervention programmes in this area was pointed to by Kilbom (1988). In 1991, a specialized project (PROSA, standing approximately for the prevention of back injuries in health care) was initiated among carers in hospitals in Stockholm (Hagberg *et al.*, 1996). In hospital work, there seemed to be a lack of systematic methods for accident investigation, of routines for systematic reporting, of analysis and feedback, and of long-term preventive strategies. PROSA's results demonstrate that interventive risk identification coupled with action-oriented feedback in cooperation with the personnel of occupational health services can promote awareness of hazards in the workplace (Menckel *et al.*, 1996). Employees' involvement in work-environment issues in general was also shown to have been furthered.

For a period of one year, the PROSA programme was implemented by physiotherapists from Stockholm County's occupational health services. A total of 130 accidents involving back injuries were investigated. The injured persons were first requested to fill in a questionnaire containing items on physical and mental load in and outside work, on previous complaints and accidents, and on the training they had received. The physiotherapists then interviewed both the injured person and his/her supervisor concerning what had caused the accident and how it might have been avoided.

Following their accident investigations, the physiotherapists fed back the information they had gathered to the workplaces where the injuries had been incurred. The information was employed as a basis for discussions of counter-measures. Everything from buying new equipment, such as easily manoeuvrable hospital beds or various kinds of equipment for the lifting of patients, to increased training and information on work techniques and the best way to move a patient was discussed.

The physiotherapists intervened reactively, taking a catalytic approach to injury prevention, and thereby offering support and encouragement so that supervisors and all categories of personnel in more than 100 hospital wards acquired improved knowledge of the risk of back injuries. Through this, personnel were enabled to identify hazards and ways of avoiding them, and come up with proposals for counter-measures.

The feedback initiated a process of development in the workplaces. Involvement in work-environment issues and working methods was promoted, as too was knowledge of and interest in back injuries and injury hazards. Even though not all proposals were implemented, various issues were raised and considered in depth.

The PROSA programme as a whole was an interventive enterprise founded on cooperation between practitioner and researcher. Various forms of intervention and cooperation grew up in the course of setting up the programme and developed further during its implementation. The principal kinds of intervention and cooperation can be summarized in accordance with the categories of our three-dimensional model as shown in Table 15.1.

It should be noted in particular that the researchers put in their greatest efforts, and had the greatest influence, during the project's initial phase, and that thereafter the physiotherapists were the driving force. It was the latter who were to survey accident hazards in the hospital wards and then feed back information to the personnel. During the project's later phases, it was intended that the initiative, and also the practical preventive work, should be taken over by the supervisors responsible, i.e. the head nurses in the wards themselves. They, in turn, were actively to utilize risk information in their preventive activities. They were also to be responsible for ensuring that the employees concerned became acquainted with this information, and were offered opportunities to discuss problems and come up with proposals for action. The dynamics of cooperative and interventive efforts to achieve prevention should not be neglected.

15.5 CONCLUSION

The three concepts, prevention, cooperation and intervention, have been treated as having a key role to play in the promotion of work that is safe and secure. During the 1980s there was, in the context of prevention, a shift in focus in the work of some occupational-injury researchers from participation in the accumulation of knowledge to an orientation towards cooperative and interventive activities adapted to and founded on local problem manifestations, conditions and resources.

To state this is, of course, not to disregard the importance of analytic research, or of models *of* injury occurrence. PROSA, for example, came into being as a result of the knowledge obtained from epidemiologic findings concerning musculosketal injuries and ergonomic (even bio-mechanical) models of how such injuries are sustained. Nevertheless, it can still be maintained that careful consideration of the interventive and cooperative elements in a model *for* injury prevention is required for the taking of action in a specific, physical and organizational, work setting.

In conclusion, it should be mentioned that although some progress has been made in finding methods for effective intervention and cooperation there is still a long way to go. This applies, for example, in the area of evaluation (see Menckel,

Table 15.1 Principal kinds of prevention, intervention and cooperation in PROSA

Prevention	Intervention	Cooperation
Primary (based on injuries incurred, i.e. reactive)	Catalytic, promoting theories and principles	Individual, leading to group/team work

1993). A variety of evaluation models and methods, each for application in its own specialized context, needs to be developed. These can then be employed to follow up the interventions that have already been made.

References

ANDERSSON, R. and MENCKEL, E. (1995) On the prevention of accidents and injuries. A comparative analysis of conceptual frameworks. *Accident Analysis and Prevention*, **27**(6), 757–768.

BLAKE, R.R. and MOUTON, J.S. (1989) *Consultation. A handbook for individual and organization development*, New York: Addison-Wesley.

ENGKVIST, I.-L., HAGBERG, M., LINDEN, A. and MALKER, B. (1992) Overexertion and occupational accidents involving the back reported by nurses. *Safety Science*, **15**, 97–108.

GOLDENHAR, L.M. and SCHULTE, P.A. (1994) Intervention research in occupational health and safety. *Journal of Occupational Medicine*, **36**(7), 763–775.

HAGBERG, M., SILVERSTEIN, B., WELLS, R., SMITH, M.J., HENDRICKS, H.W., CARYON, P. and PÉRUSSE, M. (1995) In *Work-related musculosketal disorders (WMSDs): a reference book for prevention*, eds I. Kuorinka and L. Forcier, London.

HAGBERG, M., EKENVALL, L., ENGKVIST, I.-L., KJELLBERG, K., MENCKEL, E., PERSSON, G., WIGAEUS HJELM, E. and THE PROSA STUDY GROUP (1996) Back accidents in Swedish health care. Program to counteract back injuries in the work environment of health care – for nurses, assistant nurses and nursing auxiliaries in Stockholm County. *Arbete och Hälsa*, 6 (in Swedish).

KILBOM, Å. (1988) Intervention programmes for work-related neck and upper limb disorders: strategies and evaluation. *Ergonomics*, **31**(5), 735–747.

MENCKEL, E. (1990) Intervention and cooperation: occupational health services and prevention of occupational injuries in Sweden (doctoral dissertation). *Arbete och Hälsa*, **31**, 1–48.

(1993) Evaluating and promoting change in occupational health services. Models and applications, Stockholm: Sweden's Work Environment Fund.

MENCKEL, E. and KULLINGER, B. (eds) (1996) *Fifteen years of occupational-accident research in Sweden*, Uppsala: Swedish Council for Work Life Research.

MENCKEL, E., HAGBERG, M., ENGKVIST, I., WIGEUS HJELM, E. and THE PROSA STUDY GROUP (1996) The prevention of back injuries in Swedish health care – a comparison between two models for action-oriented feedback. *Applied Ergonomics*, **28**(1), 1–7.

WESTLANDER, G. (1993) General strategies for conducting intervention studies. In *Advances in industrial ergonomics and safety V*, eds R. Nielsen and K. Jorgensen, London: Taylor & Francis.

PART SIX

Rules or trust: ensuring compliance

INTRODUCTION

How we motivate industry to adopt safe practices is a major issue. The whole focus of much of the effort in occupational health and safety is aimed at trying to understand why safety fails, trying to develop better ways of dealing with these failures and better ways of intervening to stop them occurring. All of the other parts of this book have looked at some of the most recent work and thought on many of these aspects. Unfortunately, all too often very sound safety solutions fail because they are not taken up by management or workers and incorporated into their normal work practices. Often it is not enough to have good solutions. To be implemented successfully in industry something more is needed. Traditionally this missing component for making safety solutions successful has been to make them mandatory. After all, it was reasoned, if it was a fundamentally good solution in the first place, industry or workers should use it, even if they need to be made to do so.

There are abundant examples of safety rules being used at all levels of workplace, industry and government, for example rules for the wearing of personal protective equipment in a particular workplace or rules for operating machinery that cross workplaces. There are some well-documented examples of the effectiveness of mandatory safety rules. One of the most notable was the regulation and enforcement of the wearing of seat belts in Australia and a number of other countries (Evans, 1987). This had a striking impact on the rates of wearing seat belts, reduced the severity of injuries from motor vehicle crashes, particularly head injuries, and had a marked influence in improving attitudes to the necessity for wearing seat belts.

While there are some good examples of the usefulness of safety rules, there are many other examples where safety rules have not had positive effects and have been largely ignored (e.g. Chapter 13). Part of the reason for this is that safety rules are not always based on solutions that are known to be effective. Often, not much is known about their effectiveness at all (see Part 5 for further discussion of this issue). Also, safety rules are believed to be less successful when they are not enforced. Without enforcement, the rules lose their mandatory nature and workers and managers lose their motivation to use the rules.

Largely in response to arguments like these, there has been a growing movement over the past 20 years to make industry, including individual managers and workers, more responsible for safety in Australia and in a number of other countries. Following on from the recommendations of the Robens Committee in the UK in 1972, there has been an increasing push towards self-regulation in many industries in both the UK and Australia. The emphasis in this approach has, as a result, focused on the responsibility of employers to take all reasonable steps to care for their workforce and of employees to behave in a manner compatible with safety. This approach is therefore based on trust that all parties will take their responsibility rather than on mandatory rules. The trouble with the adoption of this approach, however, is that it is being adopted even though there is little research on its effectiveness. Consequently we have little idea about whether changing the way we approach the problem of gaining compliance has improved the safety situation (Gun, 1993).

This part takes up the question of the best approach to encouraging compliance. The first two chapters take a position on the question in general, but from two vantage points, the first from a research basis, and the second from the viewpoint of the worker.

The first chapter, by Gun, Brinkman and Cox, looks at the effectiveness of a Robens-style code of practice for handling hazardous substances. This is an extension of earlier studies which showed that regulation played an important role in improving safety. Their results suggest that failing to comply with regulation would result in at least a doubling of injury rate (Gun, 1993). In contrast, industry self-regulation did not appear to produce any further lowering of injury rates (Gun, 1992). Both of these earlier studies look at regulation in general. The chapter in this part looks at the usefulness of a specific regulation for reducing injury and concludes that this code of practice does not allow for comprehensive coverage of all types of causes of injury.

The second chapter in this part, by Berger, a union representative, takes a more experiential look at the relative merits of approaches based on rules or trust. In this chapter Berger illustrates some of the occupational health and safety failures that he has encountered in his work. He argues that action to improve these situations only occurs when serious punitive action is taken against managers and supervisors who allow them to persist.

The next chapter takes up a specific example of the Australian road transport industry's attempt to develop new approaches to managing working hours. Controlling working hours is a major problem in Australia due to the large distances between destinations, particularly state capital cities. Traditionally the problem of working hours has been addressed by government regulating the number of hours that could be worked per trip and per week, and the frequency and length of breaks. Recently the state of Queensland's Department of Transport has taken the bold move of amending the working hours regulations to allow for a degree of industry self-regulation. The section by Mahon in this chapter describes the philosophy that prompted the development of the fatigue management plan approach and the actions that have been taken since it was initiated.

The next two sections put the views of different groups in the road transport industry about the fatigue management plan approach: the industry association and an operator of a small road transport company. Both sections reflect the different experiences of the groups the authors come from. Robertson, representing the industry association, is generally in favour of the self-regulatory approach but argues that to be successful it will require an attitude change by all parties, i.e. employers and employees and government. In contrast, the section by Freestone, reflecting the views of a small operator, takes quite a different view by arguing that the regulatory approach should not be judged to be a failure as it has never been implemented effectively.

This chapter does not provide a definitive answer to the question of the best way of ensuring compliance. It does, however, provide insights into the research evidence on the effectiveness of rules or trust, and the views of representatives of the major groups who would be affected by the approach used to ensure compliance with health and safety practices. The limited research that has looked at this issue in general tends to conclude in favour of the regulatory approach although further work is undoubtedly needed. Different industry groups, however, have different views of the benefits of approaches based on rules or trust. A unionist's view and the view of the operator of a small company tend to agree in favouring rule-based approaches, although for somewhat different reasons. In contrast, the arguments put by government and the industry representative in support of their new self-regulation initiative for working hours management are also compelling. It is possible that the problems with the trust-based approaches taken in the past were due to their implementation rather than a fundamental problem with the approach itself. The fatigue management plan initiative will be evaluated and the results should provide some much needed insights into the usefulness of the self-regulatory approach to ensuring compliance.

References

EVANS, L. (1987) Belted and unbelted driver accident involvement rates compared. *Journal of Safety Research*, **15**, 121–136.

GUN, R.T. (1992) Regulation of self-regulation: is Robens-style legislation a formula for success? *Journal of Occupational Health and Safety, Australia and New Zealand*, **8**, 383–388.

(1993) The role of regulations in the prevention of occupational injury. *Safety Science*, **16**, 47–66.

CHAPTER SIXTEEN

Prevention of chemical injury: an unconventional view

RICHARD T. GUN, SALLY A. BRINKMAN AND ROBERT COX

16.1 INTRODUCTION

In December 1993, Worksafe Australia adopted Model Regulations and a Code of Practice for the handling of hazardous substances in the workplace. Apart from some cavilling on the part of the Victorian Government on the question of cost, the principles underlying these measures have been generally accepted by all jurisdictions. They have been enacted in two States and are likely to be eventually adopted in some form or another in all of them. Yet it remains to be shown that they will be successful in achieving a reduction in chemical-related morbidity in a cost-effective manner.

The Model Regulations follow the principles of the so-called Robens legislation, based on the findings of a UK parliamentary committee chaired by Lord Robens. Under Robens-style legislation, which has now been adopted in most Australian jurisdictions, responsibility for ensuring workplace safety is devolved from the inspectorate to the employers themselves, who are expected to identify and control the hazards present at their own workplaces. The employer's responsibility at law is set out under the 'duty of care' provisions of the enabling legislation, which requires that the employer is required 'as far as is reasonably practicable' to ensure that employees are provided with a safe working environment.

What 'as far as is reasonably practicable' means is not defined and not definable; however, a work-caused injury or disease is *prima facie* evidence that the employer has not achieved this standard. Whilst this reversal of onus of proof can be adequate to sustain a prosecution, it does not follow that it will actually prevent injury or disease. In practice, a comparison of its performance with that of the US, which has not adopted the Robens philosophy, indicates that the latter has no demonstrable advantages over the traditional system of prescriptive regulatory control that preceded it. Figure 16.1 shows the comparative death rates in manufacturing industry in the UK and the US, and shows no particular advantage in the Robens system. Similar trends can be found by

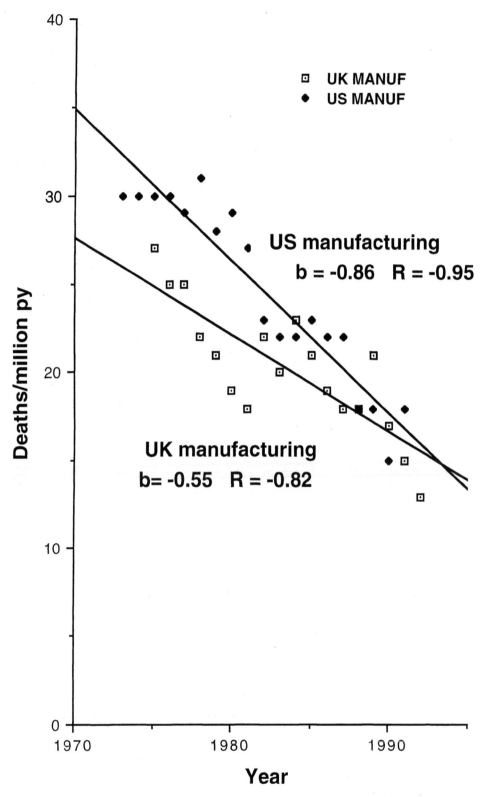

Figure 16.1 Annual death rates in manufacturing industry in the UK and the US. Source: International Labour Office, 1997.

comparing rates in the construction sector (International Labour Organization, annual publication).

Regulations enacted under Robens legislation similarly put the onus on employers to identify hazards and to develop appropriate controls. An extensive list defines which chemicals are deemed hazardous and employers are expected to assess every chemical at their workplace which is so defined. The outcome of each assessment must be documented.

16.2 LIMITATIONS OF THE PROPOSED STRATEGY

As a means of preventing adverse effects of chemicals, the proposed strategy is seriously flawed. In the first place, the requirement to carry out a formal assessment on such a large list of chemicals, which in a large number of cases present no significant risk, will mean a significant degree of non-compliance. To the extent that compliance occurs, the attention expected to be given to trivial issues is likely to devalue the currency of chemical hazards. Figure 16.2, showing a letter that arrived in the mail, is an example of this.

Another fundamental concern is that the risk assessment procedure required under the Code of Practice will fail to identify a major type of chemical-related problem: the acute exposure or chemical injury. Chemically-related morbidity is of two general types. One is disease; acute or chronic. The second type of effect is chemical injury, such as acute poisonings, chemical burns, burns from ignition of chemicals and effects of explosions. The Code of Practice requires that employers

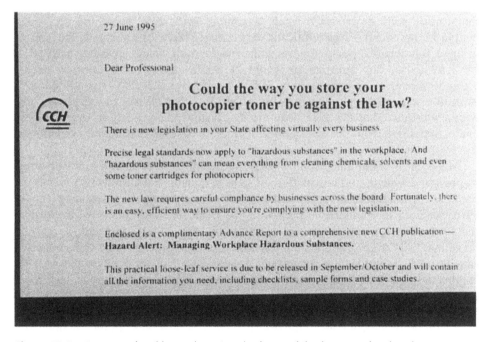

Figure 16.2 An example of how observing the letter of the law can devalue the currency of chemical hazards.

go through the processes of hazard identification, risk assessment and risk control. The risk assessment includes assessing the extent of exposure to determine whether it presents a significant risk to health, but it requires an assumption that the exposure itself is also predictable. While this assumption may be valid for assessing the long-term risk to health from the cumulative effects of constant exposure, day after day, it is not relevant to chemical injury, since this arises from an exposure that has *not* been predicted. Whereas the effects of long-term exposure can be predicted from knowledge of the effects of the chemical, chemical accidents cannot be predicted in this way since their prime cause is unpredicted exposure from administrative or engineering failure.

16.3 OTHER MEANS OF ANTICIPATING CHEMICAL INJURIES

We have been undertaking a study of the ways in which chemical accidents can be predicted and some of the cases we have studied illustrate these problems (Gun *et al.*, 1996a). One of these problems concerns a worker who sustained extensive skin burns from caustic soda when using a submersible pump to empty an electroplating tank. The clamp holding the hose on to the pump slipped and allowed the hose to separate, the result being that caustic soda was splashed on to the worker (Figures 16.3 and 16.4).

How could this accident have been foreseen? Certainly not from the risk assessment procedure prescribed in the hazardous substances Code of Practice. An assessment made at any time prior to the accident would have found no untoward exposure, no evidence of excessive caustic aerosol in the breathing zone of the workers and probably no skin contact. Nor is it likely that a material safety data sheet (MSDS) could make any difference; the MSDS no doubt says that caustic soda is corrosive when in contact with skin, something that most people know already. Nor could we expect the MSDS to anticipate the risk of using a submersible pump to evacuate caustic soda from a plating bath. If it were to do so, this information would be irrelevant to most users of caustic soda. Moreover, if we expected an MSDS to warn of this risk, it would have to warn of every other circumstance where there is a risk, clearly an impossible task with a substance that has such wide use as caustic soda (if such an attempt were made, the MSDS would be a very thick document; so thick that nobody would read it).

This point is being made somewhat pedantically to show that it is not practical to prevent accidents by anticipating all uses of caustic soda. However, it may be possible to anticipate an event such as the above example by focusing on the narrower area of submersible pumps or the even narrower area of submersible pumps used in plating shops. As a minimum, the factory inspectorate should make sure that every plating shop is make aware of this injury so that measures can be taken to prevent a recurrence. In this case the problem would best be engineered out, by an Australian Standard that requires a much more robust union between hose and pump, with a cut-out mechanism if the pressure should become excessive from a blockage or any other reason. However, even a minimal measure such as wearing protective clothing for such procedures would significantly reduce the risk.

Figure 16.3 A submersible pump being used to empty caustic soda from a tank in a plating shop.

It is important to note that this strategy focuses on a small circumscribed problem, would only prevent a particular type of accident and be only a small step forward. It is not a panacea for preventing all caustic burns but it is important to realize that there is no panacea. The Robens philosophy of merely making employers responsible for identifying hazards in their own workplace and providing general guidelines for assessment (e.g. in codes of practice) will not suffice to reduce injury rates. On the other hand, progress will result from making a series of steps that are individually small but will in time be collectively significant.

Figure 16.4   Showing the inappropriate clamp securing the hose to the submersible pump.

Admittedly these will all be solutions to problems which have occurred already but it is apparent that these apparently rare events do in fact recur. One such example was investigated early in our study (Gun, 1996a). A diver incurred fatal carbon monoxide poisoning when exhaust gas from the compressor was entrained in the diver's air intake. Of course this is a well-known hazard and to prevent this the relevant Australian Standard provides that the exhaust and air intake be separated by means of extending one or the other. This was done in this case by means of a flexible hose attached to the air intake. Unfortunately the flexible hose, which was swimming-pool hose, came into proximity to the exhaust outlet where the heat

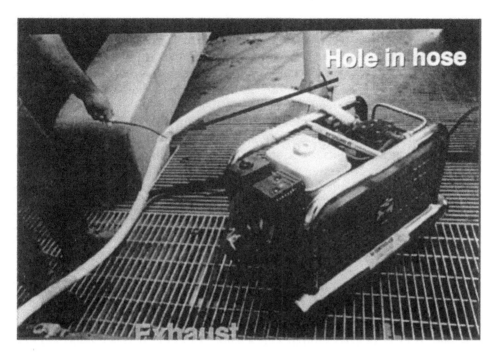

Figure 16.5 Compressor used in a diving fatality. The air intake hose came into contact with the exhaust, burning a hole in it and contaminating the diver's air supply with carbon monoxide.

burnt a hole in the hose, contaminating the diver's air supply with highly concentrated exhaust gas (Figure 16.5). Our enquiries led to the discovery of an almost identical accident two years previously in New South Wales. In that case the diving was being carried out on a floating punt, where movement caused the compressor to move, bringing the hose into contact with the exhaust.

Since diving is undertaken in so many places by small industries or individuals it is unlikely that all users could be adequately warned. Indeed in this case there was a sign on the compressor warning of the risk of the exact event which occurred. This is another case where a hardware solution is possible. The enquiry into the New South Wales accident concluded with the following recommended remedial measures:

- installation of a two-wire communication system;
- replacement of the compressor air intake hose with stainless steel;
- guarding of the compressor manifold;
- replacement of lockable caster wheels with fold-down wheels;
- compressor-securing mechanism to prevent movement.

Incorporation of these measures into an Australian Standard would prevent any further occurrences of this type of accident.

Another common occurrence where we should learn from history is carbon monoxide poisoning from the use of internal combustion engines in confined spaces.

In our small series of 23 cases we have found two people hospitalized from this situation. In one case the worker was cutting a terrazzo floor in a small room using a hand-held saw driven by a two-stroke engine. An electrically powered angle grinder was provided but the worker used the petrol-powered saw because it was lighter and hence easier to use. In the other case the worker was removing graffiti from public buildings and repainting the areas. At the time of the incident he was attaching the hose to a machine for pumping paint, driven by a petrol-driven engine located in the rear compartment of a van. He developed symptoms of carbon monoxide poisoning, requiring treatment with hyperbaric oxygen. Incidentally, both workers were teenagers recently hired by small contractors and sent to do these jobs without supervision and without adequate training in safety.

16.4 LEARNING FROM HISTORY

It is more difficult to devise a solution for episodes such as the above: certainly no hardware solution springs to mind. However, at the very least we should be providing the means by which employers and the factory inspectorate can learn from history. The ideal solution may be a database of such accidents which can be accessed with appropriate keywords, relating to variables such as the industry, process, industrial hardware and chemical. In this way employers, inspectors and others can focus on the hazards in a particular industrial process. As part of our research effort, we are examining 14 Australian accident databases to see whether they can be interrogated in this way to help predict risks by finding out what has happened before in the same situation. We have concluded that the Code of Practice for the Control of Workplace Hazardous Substances, while being potentially useful for predicting risk from continued low-level exposure, is not helpful in predicting risk from the unanticipated exposures which occur from engineering or administrative failure. Moreover, these events are possibly commoner than the occurrence of chemically caused disease. Apart from the chemically related dermatoses and respiratory diseases, diseases from the occupational use of chemicals are relatively uncommon. In a recent survey conducted in South Australia, medical specialists in haematology, neurology and nephrology were able to identify only one possible systemic disease case in three years from the occupational use of chemicals (Gun et al., 1996b). In contrast, accidents caused by chemicals are common. Table 16.1 shows the number of cases of accidents in South Australia occurring in a three-year period which led to more than five days of lost time and where the causal agent was reported as being a chemical. About one third of these accidents were manual-handling injuries (e.g. dropping a can of paint onto the foot) but this still leaves about 400 injuries (burns or poisoning) actually caused by contact with chemicals. Table 16.2 details the 113 hospital admissions to hospitals in South Australia over three years where the place of occurrence was stated to be a factory, a farm or a mine.

Table 16.1 Injuries caused by chemicals and resulting in more than five lost working days, 1988–89 to 1990–91, by agency of injury, South Australia

	Males	Females
Basic chemicals		
Industrial gases	68	14
Arsenic and arsenic compounds	2	0
Chlorine	2	0
Lead compounds	5	0
Plastic materials and synthetic resin	177	23
Radioactive material	1	0
Acid	15	4
Other basic chemicals	83	1
Total basic chemicals	353	42
Other chemical products		
Insecticides, fungicides, herbicides	3	0
Paints, varnish, solvents	15	1
Asphalt, tar, bitumen	6	0
Smoke, fire, flame	60	1
Other chemicals	108	30
Total other chemicals	192	32
Total	545	74

Source: Gun, 1996b.

Table 16.2 Occupation-related hospital separations in South Australia, 1988–89 to 1990–91, by chemical agent

	Males	Females
Carbon monoxide	21	0
Exhaust gases	6	2
Other gas or vapour	19	4
Organophosphate pesticides	4	1
Herbicides	6	0
Other pesticides	3	0
Caustic or alkali	7	2
Acids	2	0
Glues	0	2
Petroleum products and solvents	5	1
Metals	3	0
Food/plants	3	0
Lead paint	1	0
Other paint	1	0
Sulphur dioxide	1	0
Other solids or liquids	6	1
Total	88	13

Source: Gun, 1996b.

16.5 SOME RECOMMENDATIONS

We have concluded that the Code of Practice is an inadequate tool for predicting chemical injuries that follow engineering or administrative failure, which may be a more common cause of chemical morbidity than systemic disease following repeated or continuous low-level exposure. We are proposing the following additional measures to anticipate and control risks from such occurrences:

1. Establishment of a nationwide database of chemical injury, which can be accessed with appropriate keywords relating to industry, process, machinery, tools and chemicals, and will provide each industry with the particular information about what occurrences have led to serious injury in similar industrial processes. If our current study indicates that existing Australian accident data sources cannot be interrogated in this way, we shall propose a cooperative effort from which a preventive tool will emerge.

2. Development of Australian Standards, and calling them up in Regulations, wherever problems are amenable to hardware solutions.

3. Investigation of serious injuries to be followed by notification, by the inspectorate, to similar industries of the nature of the occurrence and any recommended preventive action.

As stated above, there is no panacea to injury prevention. Our research efforts suggest that there is no alternative to a painstaking analysis of individual accident scenarios and devising specific strategies to prevent them, not only in relation to chemical injuries but in manual handling and other traumatic injuries as well (Gun, 1992; Gun, 1993; Boucaut et al., 1994; Gun and Ryan, 1994).

ACKNOWLEDGEMENT

The study of chemical injuries on which this chapter is based was funded by a grant from Worksafe Australia. The final report is listed in the references (Gun, 1996).

References

BOUCAUT, R.A., GUN, R.T. and RYAN, P. (1994) An evaluation of the Risk Identification Checklist from the Manual Handling Code of Practice. *Journal of Occupational Health and Safety, Australia and New Zealand*, **10**, 205–211.

GUN, R.T. (1992) Regulation or self-regulation: is Robens-style legislation a formula for success? *Journal of Occupational Health and Safety, Australia and New Zealand*, **8**, 383–388.

(1993) The role of regulations in the prevention of occupational injury. *Safety Science*, **16**, 47–66.

GUN, R.T. and RYAN, C.F. (1994) A case-control study of possible risk factors in the causation of occupational injury. *Safety Science*, **18**, 1–13.

GUN, R.T., BRINKMAN, S.A. and COX, R. (1996a) Analysis of risk factors for chemical-related work injury, and development of a risk assessment protocol. Department of Public Health, University of Adelaide (accessible on the Internet at http:/www. health.adelaide.edu.au/ComMed/Research/Cheminj/).

GUN, R.T., LANGLEY, A.J., DUNDAS, S.J. and McCAUL, K. (1996b) *The human cost of work*, 2nd edn, Adelaide: South Australian Health Commission.

INTERNATIONAL LABOUR OFFICE (annual publication) Year book of labour statistics, Geneva: International Labour Office.

CHAPTER SEVENTEEN

The case for regulating compliance: a unionist's view

YOSSI BERGER

One of the problems with a book like this is that in terms of helping workers on a daily basis it will mean very little. It ends up being a mish mash: a bit of this, a bit of that, rules are okay sometimes and sometimes they are not, sometimes academic models and statistics are needed and sometimes they are not, depending on which sector of the industry and which hour of the day. It all seems a bit arbitrary and often over-simplified.

About three years ago, in Melbourne, a work-related death on a road construction job became Australia's first successful manslaughter case brought under the Crimes Act – not the Health and Safety Act. The company pleaded guilty and afterwards the construction industry started to provide unions with information about what sites should be investigated, machines that were not working, contractors that should be investigated and what legal clauses should be used. In other words, it was following a manslaughter case that years of effort by the unions started bearing fruit.

An inspection is due to take place of one of Australia's biggest petrochemical installations. They have recently had a government inspection, in great detail, and the report issued on the safety standard at the installation suggests that it could be a time-bomb; the entire workplace is a potential death trap. Even though for years the industry has paid lipservice to concepts like self-auditing and trust, it still takes a bad report from the regulator to get any relevant action started.

Both of these examples illustrate just how effective heavy-handed tactics are at persuading organizations to take notice of health and safety. As a contrast to the theoretical models and sophisticated analyses studied elsewhere in this book, this chapter outlines some workplace inspections to illustrate some actual workplace health and safety problems. This will also demonstrate that something more than learning, education and training is needed. As well as education and training, workplace managers need to change their behaviour to open up communication between the workplace and management. What this chapter aims to do is to demonstrate some of the 'eyes of industry'.

Example 1: The way to push a trolley (Figure 17.1)

In a large company in Australia employees have to push trolleys used to move 7 tonnes of steel some distance to the next shed. The problem is that the only way to shift the trolleys is to lie almost flat on the ground. Although it has been suggested that a small motor could be put on the front wheel to help push the trolley along, the company has placed this suggestion low on its list of priorities.

Example 2: Rooster attack

It is not nice for workers in the poultry industry to be attacked by roosters. Typically the workers bend down to pick up eggs and the roosters fly at them. The poultry industry has known about this problem for decades but has taken no effective action: no warnings or fences have been erected. In addition asthma, dust and formaldehyde exposure are major hazards in this industry.

Example 3: Forklift tyres

In a small foundry one of the forklift trucks was seen to have a very large gap in one of its front tyres, with a loose and flapping piece of rubber about to fall off. The forklift was being used to carry large sheets of very hot steel; a hazardous operation. When queried about this the manager agreed that this forklift ought to be banned.

Figure 17.1 No way to push a trolley.

Figure 17.2   The asphalt industry.

but even so he failed to take action to enforce the ban. This is an example of quality assurance and self-auditing failing in a company.

Example 4: Asphalt industry (Figure 17.2)

Figure 17.2 shows the workplace of a company that has all kinds of quality assurance systems, self-auditing and so on. In spite of this, fumes and mists are clearly visible.

   These brief examples illustrate a number of important issues about the reality of safety and the way it is managed:

- many companies realize they have had a problem for some time, for many years, but do nothing;

- many companies have extensive safety auditing systems but let safety problems persist;

- when faced with a situation that constitutes a problem many companies respond that they were just going to do something about it or simply agree that it is a problem that should not exist but still do not do anything about it; and

- many companies acknowledge some safety and health issues but ignore others; for example, some companies have extremely poorly controlled exposures to very toxic fumes and chemical mists but do not allow smoking anywhere in the plant.

The overriding point of these examples is that they exist, have existed for too long and are unlikely to change without action that makes an impact on the person who knowingly allows them to exist. Uncaring and culpable managers, who repeatedly and knowingly go on turning a blind eye to problem situations and who opportunistically exploit situations must be made to take responsibility or be made to fear goal.

It is sometimes argued that the real problem is apathy, either worker apathy or employer apathy. In practice the problem is usually fear of job loss. Many managers, and a lot of chief executives but mostly foremen and forewomen on the shopfloor use that fear of job loss for their own purposes. It is important to challenge chief executives to look at what is actually going on in their workplaces; they may well be shamed into action.

Is the answer to improving safety more rules or greater trust in workplaces? Generally a mix is required but unless the responsibility of chief executives and managers who blatantly allow poor health and safety conditions in their workplaces is increased, real improvements in safety will take a long time. No accreditation system or incentive scheme could have the same impact as sending one of these people to gaol. It is vital to change tactics to improve workplace safety.

ACKNOWLEDGEMENT

This chapter was constructed by the editors from the author's presentation at the Symposium.

CHAPTER EIGHTEEN

The case for industry self-regulation: introducing an alternative compliance system in the long distance road transport industry

18.1 NEW APPROACHES TO FATIGUE MANAGEMENT: A REGULATOR'S PERSPECTIVE

Gary Mahon

18.1.1 Background

Safety has always been, and will remain, a high priority for the Queensland Government. A safe transport system reduces the risk of injury, death and property damage for system users and assists in reducing costs. The Government has a role to ensure the safe operation of the transport system. This is generally undertaken through regulation and education programs. When developing policies in relation to safety, technical, operational and accreditation standards both the monitoring of performance and enforcement strategies are considered. Generally, any variation in safety standards should reflect different levels of risk or consequences of accidents.

The benefits of establishing and enforcing safety standards must be balanced against the costs of regulation, such as possible reduced operating efficiency. Where the community wishes to provide greater levels of safety to certain groups, for example children, to ensure that they do not bear an unacceptable level of risk, safety standards can be regulated to a higher level than would otherwise be the case. Any consideration of a need for safer transport activity must balance community expectations and the cost of meeting any potential change to safety standards.

On average, Australian state statistics over recent years suggest that fatigue is a major contributing factor in about 30% of heavy vehicle road crashes. However, safety experts believe that the incidence of driver fatigue may be under-represented and that fatigue may be a contributing factor in up to 60% of all heavy vehicle accidents and rank second only to alcohol as a cause of driver impairment in road accidents (e.g. Sweatman *et al.*, 1990).

Whilst there has been a substantial amount of research into the subject of fatigue, little if any of this research has been included in the legislative framework. Partially this can be attributed to the fact that while certain factors of fatigue may be related to accidents it is enormously difficult to demonstrate by contemporary social science methods any link between correlation and cause. However, without scientific evidence of a causal relationship between fatigue and road accidents experts can and do make various assertions without fear of contradiction. This has lead to difficult and lengthy debate on the issue around Australia.

Since approximately 1957 Australia has approached the issue of fatigue amongst heavy truck drivers from the perspective of limiting the number of hours they can drive in a day. Over the last 30 years the maximum number of daily hours has increased from 9 to 12. To administer a scheme such as this has required the use of driver log books and other associated regulatory requirements. Over time, this has translated into a reasonably cumbersome burden for both industry and the enforcement agencies. However, driving hour limits have a very tenuous link to any road safety outcomes with respect to fatigue-related crashes. The ineffective results gained from this approach have brought us to a point where clearly:

- a problem with fatigued drivers still exists;
- current driving hour limits are not managing driver fatigue well; and
- the legislation is unclear as to its purpose.

18.1.2 Environment

The traditional legislative response to road safety has been to introduce overly prescriptive, stringent and costly regulation, which is often not effectively enforced. There is a danger that the drafting of laws designed to promote equality and safety will achieve the opposite. A case in point is that of driving hour limits. This type of regulation gives comfort to the community that fatigue is controlled. Unfortunately, this perception is misguided. The driving hours regulatory regime, by its nature, does not allow drivers to get adequate sleep when they may need it. Rather, drivers are constrained to take rest on the basis of a standard formula specified by the regulations. In addition, the regulatory regime also fails to consider other needs such as eating habits, recreation time, quality rest or the effects of inability to fall asleep at the beginning of a consecutive rest period as required by the formula in the legislation.

Perhaps the bigger problem, however, is the role played by the regulated driving hours regime in training operators to think about driver fatigue. Having the regime in place encouraged operators to not only think that any driver should be capable of 12 hours of driving a day, 6 days a week but, further, that if they only extracted this amount of driving hours out of each driver per week then they were meeting their fatigue management responsibilities irrespective of the state of health of that driver. Meeting the regulatory requirements rather than managing fatigue was seen as the appropriate outcome.

The challenge, therefore, was to bring an appropriate level of outcome orientation to the fatigue management legislation. The aim was to ensure that operators and drivers understood that the objective was to make certain 'that the driver of any heavy vehicle was always in a fit state of health and well-being to safely drive that vehicle'. This performance standard is fundamental in the approach to legislation in this area in the future.

Our thinking behind the introduction of the performance-based standard is as follows. Introducing a standard and providing the flexibility to meet its outcome means that credibility is earned for the regime and, consequently, operators as well as drivers will be more likely to want to meet the standard voluntarily because they respect and understand its relevance. This voluntary system is likely to attract higher compliance without the necessity for direct enforcement intervention and will simplify the compliance task. Further, for the regulator, it provides a reasonable basis to apply a targeted enforcement strategy. Our experience suggests that approximately 80% of the population will try to do the right thing, so long as they see strong action being taken about the 20% who do not. A voluntary system such as the one we envisage means that the 20% become more clearly identified.

18.1.3 Cultural change

The move from a prescriptive focus to a performance basis in the road transport industry constitutes a major cultural shift. In any large-scale cultural change the critical element is to ensure that the purpose of the change is clear and that the context of the change in your vision of the future can be justified and explained to the ordinary person. The relative success of the change programme will depend largely on its first impression credibility, which will be judged against such criteria as:

- a maker not a breaker;
- a launching pad not a trap; and
- a beginning not an end.

To achieve cultural change in the road transport industry, our approach is built around four main steps:

- *Picture*: Ensure that you clearly enunciate what it is that you are changing and why, and illustrate the benefits for individuals and the community. The value of fostering altruism is not to be underestimated in terms of community responsibility for any industry in the modern business environment.

- *Permission*: When the legislative framework allows for the change, you are more likely to change the given community behaviour as it is seen to not only have government approval but also community support.

- *Protection*: Whilst legislation does not give any protection for negligence in its lawful meaning, if the inclusion of enablers of the behaviour change within the law are given support by the transport agency, you will more readily encourage your 'product champions' to take the lead in the change.

- *Process*: The behaviour change needs to incorporate a set of processes that the industry can follow to assist in its management. The first steps in the change process should not be enforced by government. They should be undertaken voluntarily by industry.

18.1.4 Legislation

The existing legislation in Queensland at the time of our review (1992–93) was typical of national and international practice. It was a driving limits approach that contained a prescriptive formula for the hours that could be driven and the rest taken. This was the only method that was used by government for safety and efficiency purposes. This formal approach has an outcome of high standardization with the consequence of low discretion.

The redesign of the legislation needed to ensure that a performance-based approach was taken that was underpinned by meaningfulness for the user and clear responsibilities for outcomes from the owner and/or driver, depending on their respective roles. In terms of the elements that the new legislation needed to include, the following were applied:

- *A performance standard*: a clear explanation of the performance the law requires as a minimum.

- *Procedures*: the procedural steps to achieve the performance standard.

- *Decision principles*: the warrants that the administrator will require to make decisions in regard to performance.

- *Prerogatives*: the rights of the Crown in making determinations about the administration of the particular piece of legislation.

- *Sanction structure*: the punishments for non-compliance.

- *Flexibility*: to accommodate changing conditions and continuous improvement.

18.1.5 The intervention

On 1 March 1994 Queensland introduced a completely new set of statutes replacing the previous driving hours regulations. These new requirements are now called fatigue management regulations. The title clearly illustrates their purpose and they show considerable promise of being a very powerful tool to successfully move the industry to the correct focus of management and responsibility rather than just being subjected to enforcement.

The legislation contains a safety net with a set of prescriptive requirements for those operators who do not volunteer their participation. In other words, if an operator is not prepared to do anything to manage fatigue, then as a minimum community standard they must at least adhere to driving hour limits. These requirements include:

- a maximum of 12 hours of driving in a day, calculated from midnight to midnight;

- a maximum of 9 hours of cumulative rest in any 24 hours of their journey of which 6 hours must be consecutive;

- driving not more than 72 hours in 7 days;

- driving not more than 5 hours consecutively;

- at least a 30-minute break after 5 hours of driving.

Also included are regulations relating to the log book completion, timing of that completion and surrender of logbooks, as well as a whole variety of other prescriptive requirements.

As a subset of these prescriptive requirements, exemptions and variations are also included. It must be stressed that exemptions and variations only relate to the prescriptive regulations. These could apply, for instance, to a particular shire during a period when all vehicles hauling grain from farm to silo require an extra hour's driving time or maybe exemption from completing their logbooks because they have an alternative method for keeping records. Individual requirements can be catered for on the right grounds. These grounds are all set out in the Queensland Department of Transport fatigue management exemption policy.

However, the most important part of this legislation is the fatigue management program (FMP) option. An FMP is a programme that manages all the fatigue risk factors that can affect the heavy vehicle driver's fitness to work. It consists of documented assurance systems, policies, procedures and records, which show management and evaluation systems are in place to ensure compliance with agreed fatigue management standards. If a company comes to the Queensland Department of Transport and can provide a fatigue management programme that demonstrates and guarantees that it will ensure the health, fitness and safety of any person driving a heavy vehicle, they will be allowed to step outside prescriptive regulations. The opportunity is there for a company to have flexibility of operations within their own management environment where they are not obliged in any way, shape or form to comply with any aspect of the prescriptive regime. The test is that they must demonstrate a much better outcome. The operator has to outline to the Department how they would manage fatigue in their fleet.

What this is doing is shifting the accountability from the driver to the owner. The owner has everything to lose because they can now tender for a contract within their own requirements so long as they are managing fatigue and can ensure fatigue-related accidents do not occur. They have a real opportunity for tangible benefit for themselves. If they go back to the prescriptive regime it can be more difficult for them to function. Most operators, if put back to the prescriptive regime, are likely to find difficulty because what they had before is bound to have been a much better management arrangement than the prescriptive regulations. What we must remember is that this option is based on meeting a better performance standard. It is not a question of demonstrating how good we are at forcing people to comply with systems we invented (e.g. the log book) but with the actual reduction of fatigue-related accidents.

Consequently, the fatigue management programme fits into the process of managing the inputs better. We cannot be conclusive that the fatigue management programme will reduce the total of fatigue-related accidents but nevertheless we know what the inputs are and what stresses, hours and other matters make a person fatigued.

The output is that a certain amount of fatigue is generated in the system, from which the outcome is fatigue-related accidents. If a process is put in place to manage and countermand those influences and reduce the amount of inputs in the system, it is reasonable to conclude that on a balance of probabilities the amount of fatigue in the system will be reduced and thereby the number of fatigue-related accidents will be reduced. This process will make the cultural shift away from being dependant on the number of hours driven, i.e. 'that will do and I've met my responsibilities', to a situation where operators and drivers understand that their responsibility is to ensure that they are managing fatigue.

18.1.6 Fatigue management programme

The first requirement of the FMP is that the owner must have a management system in place that will withstand audit scrutiny at any time. An FMP system makes the owner accountable for the final outcome, i.e. the owner will have everything to lose and, because of this, will be more likely to bring some credibility and integrity to the information being provided. The agreement the owner enters into with the Department of Transport outlines the 'whats', that is what the performance standards are against which the company will be measured. It is up to the individual owner to show how these requirements will be met in the management programme within their system to demonstrate to the Department how they are going to prove they are doing what they say they are doing.

The Queensland Department of Transport is currently conducting a pilot project and over the next 12 months there will be over 25 pilot operators throughout the country. Every state that administers a driving hours' scheme is participating and the project has National Road Transport Commission and Federal Government endorsement. Professionals within the health arena, such as ergonomists and other scientists, have significantly assisted in the development of this programme. Pilot operators will soon be underway with the objective of producing a report by July 1998, so that later that year guidelines can be published for a programme for any person that wishes to participate, whether they have one truck or 1000 trucks.

Accreditation standards for the programme are essential as they are the benchmarks against which the credibility of compliance will be judged. These standards are going to have to withstand substantial scrutiny to ensure we are comfortable to allow the operator to function outside the prescriptive regulations. This assurance is essentially about the types of records that can be examined during audit that demonstrate that operators are doing what they say they are doing.

There are many issues that the Department is now aware of, from a considerable body of published research, concerning factors that contribute to risk in terms of the potential of fatigue to cause accidents (e.g. Harris and Mackie, 1972; Harris, 1977; Wertheim, 1978; Lisper *et al.*, 1979; Fuller, 1984; Lisper *et al.*, 1986; Jones and Stein, 1987; Hertz, 1988; Hawarth *et al.*, 1989; Williamson *et al.*, 1992; Williamson *et al.*, 1994; Feyer *et al.*, 1995; Feyer and Williamson, 1996). It is for these types of issues that the fatigue management programme for each operator will need to demonstrate appropriate counter-measures or interventions to reduce risk to acceptable levels. Some of the issues are:

- Fatigue-related accidents are more common on rural highways than on urban roads. One reason for this is that average trip lengths are likely to be longer on these roads and inattention and drowsiness are brought on by the constant speeds and monotony.
- Twice as many crashes occur in the second half of a trip as in the first half.
- The crash rate is at its most inflated during the fifth and sixth hour and diminishes somewhat thereafter.

- The crash rate is 2.5 to 3 times higher after 14 hours or more of work than for working periods of 10 hours or less.

- The crash rate for drivers of articulated vehicles who have driven for more than 8 hours is double that of drivers who have driven for less than 8 hours.

- Most crashes that involve drivers falling asleep at the wheel occur at night.

- Twice as many crashes occur between midnight and 8 a.m. as in the rest of the day and about half of single-vehicle crashes occur in the early morning hours.

- The effects of prolonged driving depend in part on when that prolonged driving takes place rather than simply on its actual duration.

- Drivers who have completed a day of physical work drive less competently and are more likely to fall asleep than are drivers who have completed a day of mental work or who have rested that day. Therefore the effect of prior activity in hastening the onset of fatigue should be accounted for.

- The type of activity during rest breaks is important; walking or eating improve rest quality.

- The characteristics and effects of driving in remote areas increase fatigue (more monotonous, dozing off, leaving the roadway, drifting across the road, etc.) but there is a reduced risk of accident because there are fewer intersections, vehicles, pedestrians, roadside poles, trees, etc.

- Drivers operating on an irregular schedule suffer greater subjective fatigue, physiological stress and performance degradation than drivers who work a similar number of hours on a regular schedule.

- Accumulating 8 hours sleep in two 4-hour periods results in less rapid eye movement sleep than sleeping 8 hours at a stretch. When the body has a shortage of rapid eye movement sleep the person may still feel drowsy even though they have accumulated sufficient hours of sleep.

These are factors that we know about and they are the types of issues that we are looking for answers to, along with counter-measures that the manager is able to produce. If a driver is put on the road for 15 hours, what is going to be done to balance that and what counter-measure is going to be put in place to ensure that this is not a problem. At the end of the day there is a lot more at risk for the manager. It is certain that managers are not going to want to go back to the prescriptive regime because the fatigue management programme is a commercially competitive opportunity at a higher safety standard. That standard is not negotiable and cannot be compromised. The programme is comprised of the following elements:

- *Scheduling*: scheduling of all trips must incorporate the fatigue management measures required to meet the freight task and also provide drivers with the flexibility to reschedule driving and rest periods, if required.

- *Rostering*: rostering systems must be in place that assign drivers to tasks in accordance with their recent work history, ability, welfare and preference, where appropriate.

- *Time at work*:
  - drivers must have the ability and opportunity to effectively manage their driving and non-driving work time in a way that allows them to take measures to combat the effects and onset of fatigue.

- the operator must not allow or cause a driver to work for periods that may endanger the safe operation of the vehicle and expose the driver, other road users and the environment to unacceptable levels of risk.
- the operator must demonstrate that scheduling and rostering policies and techniques are being practiced and ensure that accurate records of each driver's daily time at work and rest at work activities are kept.

- *Rest at work*:
  - the operator must allow a driver a minimum rest time to combat the effects and onset of fatigue after a continuous period of driving time.
  - the operator must ensure that a driver can reschedule rest and driving periods, if required.

- *Readiness for duty*: the operator must require all drivers to be in a fit state to safely operate a vehicle and perform other necessary duties as required.

- *Time not at work*: the operator must ensure a driver has sufficient continuous hours of time not at work to recover from the effects of fatigue caused by a period of time at work and extended periods of time at work.

- *Health*: the operator must ensure a health management and screening system is in place to best prevent and combat the onset and effects of fatigue and to address as a minimum such factors as medical history, sleep disorders, diet and substance abuse and provide preventative and remedial measures to assist drivers with the management of their health.

- *Management practices*: management practices must ensure all drivers are suited to the freight task and that open lines of communication are fostered between management and drivers on matters that may enhance the safe operation of the business.

- *Workplace conditions*: workplace conditions must provide environments that assist in the prevention of fatigue.

- *Vehicle safety and road access requirements*: the operator must ensure that all vehicles owned and/or operated by them are safe to use on the road and that these vehicles do not expose other road users to unacceptable risk.

- *Driver road use requirements*: the operator must ensure that drivers are competent, licensed and authorized to drive the applicable category of vehicle safely on the road and in accordance with prescribed driving standards.

- *Training and education*: the operator must identify the fatigue management training and education needs of all employees and ensure that every staff member, including managers, is provided with training and education on the management of fatigue and the operator's fatigue management programme.

- *Documented policy and procedures*: the operator must prepare, implement and maintain documented policies and procedures that ensure the effective management, performance and verification of the fatigue management and accreditation requirements of the operator's FMP.

- *Responsibilities*: the operator must assign and document the responsibilities and authorities of all positions involved in the management and operation of their FMP.

- *Management of non-compliance and corrective action*: the operator must ensure all FMP non-compliance occurrences are reported and corrective action and preventative measures are taken in accordance with the level of risk identified.

- *Records*: the operator must ensure the identification, collection, storage and maintenance of records that demonstrate compliance with FMP standards; records are to be kept for a minimum of three years.

- *Documentation controls*: the operator must operate a system to authorize, review and control all documents, including manuals, procedures, reference materials and legislation, required for the administration of a FMP.

- *Internal audits*: the operator must have an internal audit system to verify that all FMP activities and record-keeping procedures comply with directions given in the relevant policy, procedures and instructions and that any corrective action required has been undertaken.

- *Identification of participants*: the operator must ensure that all persons and vehicles participating in the operator's FMP are identified and their details and status as participants are recorded.

- *Risk management*: the operator's FMP must demonstrate how the operator's fatigue management system and procedures can effectively manage any fatigue risk caused by the operator's freight task to ensure that it will not expose the operator's drivers, other road users and the environment to unacceptable levels of risk.

These standards will need to be developed to a sufficient extent that accreditation principles can be applied. In signing the agreement terms with its accompanying conditions the operators will be binding themselves to a comprehensive management arrangement to meet fatigue-based performance outcomes.

18.1.7 Conclusion

In the pursuit of best value and environmentally sustainable transport services the Queensland Government encourages continuous improvement in the safety and efficiency of the freight transport task. To achieve this the Government must work together in partnership with industry. However, in progressing major changes within the industry it is necessary to focus on the development and management of the individual, be they driver, supervisor or manager. Well-trained, well-informed and well-managed staff are more important than all the new policy or technology that can be mustered.

This philosophy underpins the development of the fatigue management programme, an initiative by the Queensland Department of Transport to move towards performance-based legislation to manage a major occupational hazard in the long-distance road transport industry. The programme targets the development and implementation of management training, schedules and education programmes that focus on fatigue and outline the need for drivers to acquire amounts of quality sleep, develop strategies for avoiding sleep loss and consider the behavioural and physiological consequences of tiredness. This will enhance awareness that sleep can occur suddenly and without warning to all drivers regardless of their age or experience and that fatigue has a serious effect on a driver's work performance and safety. Successful management of driver fatigue involves a cooperative approach between management and drivers to long-distance driving operations. It is about balancing the fatigue levels of each driver and providing the appropriate counter-measures to alleviate the impact or onset of fatigue.

References

FEYER, A.-M. and WILLIAMSON, A.M. (1996) Work and rest in the long distance road transport industry in Australia. *Work and Stress*, **9**(2/3), 198–205.

FEYER, A.-M., WILLIAMSON, A.M. and FRISWELL, R. (1995) *Strategies to combat fatigue in the long distance road transport industry. Stage 2: Evaluation of two-up operations*, Report CR 158, Canberra: Federal Office of Road Safety.

FULLER, R.G.C. (1984) Prolonged driving in convoy: the truck driver's experience. *Accident Analysis and Prevention*, **16**, 371–382.

HARRIS, W. (1977) Fatigue, circadian rhythm, and truck accidents. In *Vigilance: theory, operational performance, and physiological correlates*, ed. R.R. Mackie, pp. 133–146, New York: Plenum Press.

HARRIS, W. and MACKIE, R.R. (1972) *A study of the relationships among fatigue, hours of service, and safety of operations of truck and bus drivers*, Final report BMCS-RD-1727-2, Goleta, California: Human Factors Research, Inc.

HAWORTH, N.L., HEFFERNAN, C.J. and HORNE, E.J. (1989) *Fatigue in truck crashes*, Report No 3, Melbourne: Monash University Crash Research Centre.

HERTZ, R. (1988) Tractor–trailer driver fatality: the role of non-consecutive rest in a sleeper berth. *Accident Analysis and Prevention*, **20**, 431–439.

JONES, I.S. and STEIN, H.S. (1987) *Effect of driver hours of service on tractor–trailer crash involvement*, Arlington, VA: Insurance Institute for Highway Safety.

LISPER, H.-O., ERIKSSON, B., FAGERSTROM, K.-O. and LINDHOLM, J. (1979) Diurnal variation in subsidiary reaction time in a long-term driving task. *Accident Analysis and Prevention*, **11**, 1–5.

LISPER, H.-O., LAURELL, H. and VAN LOON, J. (1986) Relation between time to falling asleep behind the wheel on a closed track and changes in subsidiary reaction time during prolonged driving on a motorway. *Ergonomics*, **29**(3), 445–453.

SWEATMAN, P.F., OGDEN, K.J., HAWORTH, N., VULCAN, A.P. and PEARSON, R.A. (1990) *NSW heavy vehicle crash study final technical report*, Report No CR 92, Canberra: Federal Office of Road Safety.

WERTHEIM, A.M. (1978) Explaining highway hypnosis: experimental evidence for the role of eye movements. *Accident Analysis and Prevention*, **10**, 111–129.

WILLIAMSON, A.M., FEYER, A.-M., COUMARELOS, C. and JENKINS, A. (1992) *Strategies to combat fatigue. Stage 1: The industry perspective*, Report CR 108, Canberra: Federal Office of Road Safety.

WILLIAMSON, A.M., FEYER, A.-M., FRISWELL, R. and LESLIE, D. (1994) *Strategies to combat fatigue in the long distance road transport industry. Stage 2: Evaluation of alternative work practices*, Report CR 144, Canberra: Federal Office of Road Safety.

18.2 ROAD-TRANSPORTATION: INDUSTRY PERSPECTIVE

Denis Robertson

The Road Transport Forum (RTF) is the peak national body of the Australian road transport industry, an industry that employs 400 000 people and turns over about

$A15 billion every year. The main goals of the RTF are to secure an operating environment that promotes safe, efficient road transport services, free of excessive regulation and government taxes and charges, and to promote improved standards of safety and professionalism throughout the road transport industry.

To achieve these improved standards, the RTF is implementing a ground-breaking industry accreditation project. We believe that it is one of the most impressive accreditation projects ever undertaken. The industry accreditation project started in 1990 and now involves around 3000 drivers in 300 companies, which form the group of participants that we call the Team 2000. The key to the industry accreditation project's success, we believe, is the development of a realistic set of standards that are acceptable both to operators and governments.

The responsibility for upholding these standards within the accredited companies is placed on the shoulders of each and every company member, from the driver to the manager to the chief executive. This has really changed the focus from being mainly on the drivers to include everyone in the company. The author runs a transport company that is part of the Team 2000 group of accredited companies. In the past in this organization, we often said to drivers, 'If you feel tired, you stop and have a sleep'. Now we still say that, but we also make much more effort to ensure that drivers do stop and have a rest or whatever it is that they need. The result has been no significant accidents for over three and a half years. We are proud of this, especially considering that our trucks cover around 20 000 km every day.

The way such a good safety record has been achieved has been through participating in the Team 2000 programme of implementing standards in the critical areas of drivers' health, driver training, vehicle maintenance and management. Management in particular has a very large role to play. Occupational health and safety is one of the critical areas in which the accreditation approach can greatly assist in improving standards of safety and professionalism and generate savings for companies, the community and governments. The industry accreditation programme can act to reduced the incidence of occupational injuries by requiring industry participants to implement procedures that maintain standards of safety and where necessary take remedial action to correct unsafe practices.

The Team 2000 standards are being designed and implemented in close partnership with governments and their agencies. It is our view that without this continued liaison between industry and government, there is little hope of successfully improving the situation beyond what was possible under the old regulatory compliance schemes. This liaison has required trust on the part of government that the road transport industry is prepared to improve and to take standard setting and enforcement into their own hands.

We recognize that there are great benefits for government in the accreditation approach. By allowing transport companies that endeavour to do the right thing to participate in the industry accreditation programmes, government can concentrate on targeting companies that are not doing the right thing. This means that the task of regulating becomes much less difficult, especially where there are scarce resources for policing. However, government needs to maintain a realistic level of funding from the savings it makes through more effective targeting of enforcement resources to ensure the long-term future of industry accreditation.

There is considerable scope for savings also for the road transport industry and individual companies. In the state of Victoria alone, for example, during 1993–94, road transport companies registered with WorkCover, the workers' compensation authority, paid an average of $A13 500 in workers' compensation costs for each $A1 million of industry remuneration. Total days compensated for over the same period were the equivalent of 200 working years lost. These figures also give some idea of the scope for improvement through schemes such as the accreditation project run by the RTF.

In conclusion, there are three points that should be emphasized. First, the road transport industry must be consulted as a full partner in any regulatory development because they know best how industry operates. This should not be one-sided because regulators and industry need to work together, but the only way that there is ever going to be any respect for legislation is when people see that the legislation has some logic. When people have some respect for legislation, they will adhere to it.

Second, the transport industry is part of the nation's infrastructure. It is responsible for moving raw materials, primary products and finished articles all around the country. It is important therefore to keep the industry safe and viable. Third, the industry is currently implementing a behavioural change, but it must be realized that the move to industry accreditation needs a cultural change as well.

Lastly, the industry will accept the change to self-regulation if they see it as relevant. The industry, however, needs support from governments and their agencies and this includes contributing to the cost of developing self-regulating programmes. The road transport industry is very anxious and very committed to do what it can to bring improvements, but it must have the help of government to do this.

18.3 DRIVING HOURS REGULATIONS:
A TRANSPORT OPERATOR'S PERSPECTIVE

Paul Freestone

The author has worked in the road transport industry since 1966, starting off with an old B61 Mactruck. The task in those days was to head down to Dandenong, an industrial suburb of Melbourne, put a full load of product from the local motor car building plant on the truck and head up to Townsville in Queensland, a trip of about 3000 km. Around 500 km into the trip we would make a 200 km diversion where the load would be removed, a full load of wine loaded underneath and the load of vehicle products put back on top. Of course, this meant that the load was then overweight and too high. It took two weeks to get to Townsville and a month to do the round trip. We probably broke every regulation on the way to Townsville, and most of them on the way back. In 1966, this was thoroughly acceptable. Now, in 1996, many operators in the road transport industry would do that task in around four days. This is clearly a real improvement in travel times, although to make this improvement every law in the book is probably still broken. There is no umpire to prevent this and although companies seem to be more efficient the truth is that they have just developed better ways of making the person behind the wheel work harder.

In Australia, because there are very long distances between major transport destinations, working hours have always been a problem. The task for many drivers, however, involves more than a 12-hour trip from point A to point B. Drivers also need to do the distribution of the load. Take, for example, the driver who reaches his destination in a state capital city after a legal 12-hour trip. When he arrives at the outer suburb of the city, he merely puts in his log book that he has arrived in that suburb. He then proceeds to write 'off-duty' for the day but, instead of resting, he loads and unloads the freight. When the driver gets back to the suburb at the beginning of his return journey, he merely writes 'back on duty'. That driver has just worked eight or ten hours on top of his original trip and will do another 12 hours to return home. Sometimes drivers can do six to ten deliveries. Sometimes they might be able to load and unload within two or three hours, but with very large trucks and double trailers, this is becoming less and less common. This example is a reality today; that is how it happens in probably 90% of the road transport industry. The industry condones it because that is the way they have always been allowed to operate. This is not the fault of the truck drivers, the companies or the enforcement authorities; it has simply been allowed to happen.

Where does the industry go from here? New laws allow any kind of new regulation but if they are not enforced, they are a waste of time. Moves such as industry self-regulation and fatigue management programmes are good but if there is no one out there to police what drivers and companies are doing, the problems will still exist.

When working hours are not enforced drivers and companies have a major problem. Drivers can routinely work 18 hours per day and the only time they get a proper break is when they run out of freight or the truck breaks down. This has obvious repercussions for safety on the road and leads to an increase in the problem of the use of stay-awake drugs in the road transport industry. For companies, no real enforcement of working hours means that there is unfair competition within the industry; there is an unlevel playing field. In the author's small company, regulations are enforced by the use of a dock to dock operation, provision of accommodation in all states in which the company operates and not allowing drivers to load and unload the vehicles. The good accident record of the company proves that this is the way to get the job done properly. However, this company is in the minority.

The problems affecting the occupational health and safety of interstate drivers could be eliminated overnight if the current laws that are set out in the log book were enforced. The defined driving hours provide guidance about the work procedures that should be followed. Overworking drivers is a fairly common feature of the industry. Employees are expected to unload and load during the day, drive all night, then unload and load the next day. If the driver is lucky, he manages to get a few hours sleep along the way, only to repeat this process many times. Many companies see nothing wrong with this process, after all, it is the way the industry has always operated.

There is an alarming number of single-vehicle accidents. Why should these figures be on the increase when roads are improving and truck speeds have been reduced? The answer to this is obviously fatigue. Governments need to look closely at the different types of transport operations. On the surface they may seem to be similar but they operate with completely different work practices. By adhering to proper working hours, as set out in the log book, maintaining equipment and ensuring that drivers gain sufficient rest away from the truck, the problems of fatigue could be reduced and consequently the number of accidents in the road transport industry could also be reduced.

PART SEVEN

The compensation system in Australia: help or hindrance?

INTRODUCTION

In most industrialized nations, workers' compensation was introduced in the late nineteenth and early twentieth centuries. In Australia, following the UK and New Zealand precedents, legislation was introduced with the central tenet being the relationship between loss of income and injury arising during the course of work (Blackmore, 1993). All Australian states had adopted such legislation by 1914. Prior to this legislation, workers injured during the course of their employment had been able to make claims for damages under common law, which recognized that injury could result from work. However, the question of liability in such claims was problematic because they were dependent on the worker being able to establish a specific relationship between the current work and the injury before the current employer could be held responsible (Kenny, 1994). The legislative changes at the turn of the century were the first laws that dealt specifically with the working environment, formally recognizing that injury could be the consequence of work and not necessarily the negligence of the worker. These laws made it mandatory for employers to take out insurance covering employees against income losses and medical expenses resulting from injuries arising in the course of work (Craigie *et al.*, 1986) and provided a right, without question of fault, to a package of benefits which accrued from employment (Kenny, 1994). Insurance was also needed to cover liabilities under common law, since employees could still claim damages under common law if they considered the compensation provision inadequate or if the injury was considered to be due to negligence or breach of statutory duty.

The legislation in Australia remained largely unchanged until the 1980s when mounting pressure for reform arose out of the spiralling costs of premiums for industry, the spiralling costs of payouts for insurers and increasing awareness that the potential of the system to provide an incentive to employers to prevent injury was not being realized (Kenny, 1994). In particular, it became increasingly recognized that workers' compensation insurance *per se* is not a preventive measure because it distributes the cost of injury equally across all employers (Hopkins, 1994). Over the last decade, most states have introduced, or are considering introducing, premium incentive schemes similar to those operating in other parts of the world, for example in the US and Canada. Under such schemes, premiums are weighted to reflect an employer's claims record.

Advocates of such systems of weighting premiums in relation to claims experience suggest that the principle behind this system is a sound one. Employers have an economic incentive to minimize occupational injury and disease, with better performance being rewarded. To data, however, few studies have evaluated empirically the impact of incentive schemes on safety performance, either internationally or in Australia. Hopkins (1994), reviewing the evidence that does exist, suggested that both internationally and in Australia the evidence is inconclusive. Some studies do suggest that premium schemes have a beneficial safety effect, but the published data do not demonstrate the effect unambiguously.

There has been considerable debate about the potential limitations of the system. The focus of the chapters in this part is the potential for the key generic aspects of the compensation system, as it currently operates in Australia, to achieve gains in injury prevention. Both chapters provide discussion of the various ways in which a system based on weighted insurance premiums should work, and how it may fall short of achieving its potential for accident and injury prevention. The chapter by Leigh also explores the limitations of claims at common law, the other major recourse currently available in Australia to employees injured at work. Leigh considers the potential role for such claims as economic incentives for employers to institute safer work systems and considers their limitations. His chapter concludes that economic incentives alone are unlikely to be adequate and that there remains a role for criminal sanctions.

The chapter by Hopkins takes up the issue of the role of enforcement in improving safety performance in rather more detail. Essentially, the weighted insurance premium system can be seen as one encouraging more effective self-regulation of safety by employers in exchange for the economic advantages offered by lower premiums. Hopkins argues that the compensation system should not be seen as the only legislative incentive for companies to improve their performance. Rather it should be seen as part of a wider range of incentives. The chapter argues for the important role that inspectorates have to play and provides evidence concerning the impact of inspectorates for reducing injury and improving safety performance. These issues are discussed further in Part 6, which specifically takes up discussion of the role for regulations and prescriptive regimes in preventing occupational injury, as compared with alternative mechanisms such as performance-based systems.

Two key issues emerge from the chapters in this part. First, the workers' compensation system in Australia aims to provide economic incentives to employers to prevent injury through weighted premiums and to a more limited extent through claims at common law. However, the nexus between these incentives and injury reduction may not be clear cut or direct. Second, it is clear that better empirical evaluation is needed of the impact of the compensation system on injury rates and patterns. While there are convincing *a priori* reasons supporting the potential benefits of economic imperatives, their actual effect on such phenomena as claims suppression or differential injury prevention requires further research.

References

BLACKMORE, K. (1993) Law, medicine, and the compensation debate: Part 1. *Journal of Occupational Health and Safety, Australia and New Zealand*, 9(1), 59–63.

CRAIGIE, R., CUMPSTON, R. and SAMS, D. (1986) Accident compensation reform. *The Australian Economic Review*, 3rd Quarter, *Issues in Social Economics*, 9–30.

HOPKINS, A. (1994) The impact of workers' compensation premium incentives on health and safety. *Journal of Occupational Health and Safety, Australia and New Zealand*, **10**(2), 129–136.

KENNY, D. (1994) The relationship between workers' compensation and occupational rehabilitation: an historical perspective. *Journal of Occupational Health and Safety, Australia and New Zealand*, **10**(2), 157–164.

CHAPTER NINETEEN

Workers' compensation and common law: how the civil legal system discourages occupational injury prevention

JAMES LEIGH

19.1 INTRODUCTION

After the eclipse of the paternalistic agrarian economy by the industrial revolution in the late eighteenth and early nineteenth centuries, the English courts created a pattern of industrial law that reflected the postulates of individualism as propounded by Adam Smith. If each person sought solely to act only for his individual economic gain, through 'freely' negotiated contracts, an 'invisible hand' would somehow ensure that free-market forces also worked for the common good of society. However, this theory conveniently ignored both the inequality in bargaining power between management and employee and the measure of economic compulsion that left the employee no realistic choice between acceptance of the conditions of work offered to him and starvation or equally hazardous employment elsewhere.

With a view to encouraging and subsidizing rapidly developing capitalistic enterprise, the standard of protection conceded to employees was the minimum employers were capable of affording according to the rather unexacting standards of the time. By means of reading fictitiously implied terms into the contract of employment, the courts denied redress for injury from the inherent dangers of work and also excused the employer (the master in the master–servant contract of employment) from vicarious liability for the negligence of fellow-servants (the doctrine of common employment). The employee was regarded as voluntarily assuming these risks, incident to the supposed free and equal bargaining position in the employment contract. These doctrines, plus the defence of contributory negligence, ensured that the great majority of the huge numbers of industrial accidents went uncompensated at common law in either tort or contract (Fleming, 1983).

In the latter part of the nineteenth century, this position began to be somewhat moderated by the introduction of workers' compensation schemes, the first, that of Germany in 1883, actually not being in a common law country and really introduced by Bismarck to undermine socialist unrest among the workers' movements. As Bismarck said 'The social insecurity of the worker is the real cause of their being a peril to the state' (Elling, 1986). The German schemes were jointly funded by employers and the state so the liability was distributed further than workers' compensation systems in Australia.

In the USA, state workers' compensation schemes came into being after 1910 when the National Association of Manufacturers and the National Civic Federation, organizations dominated by large corporate members, began to lobby for such measures. While these measures gave some aid to injured workers (with little or no recognition of work-related disease) their main purpose was to protect employers from unpredictable and increasingly large and successful common law claims. Following adoption of workers' compensation laws, the worker could no longer sue his or her employer, although third party suits were still allowed (e.g. asbestos). The employer thus had a predictable 'cost of doing business' with possibly lower outlays than continuing law suits may have entailed. The benefits were uniformly low and nowhere near the value of lost wages.

The doctrine of common employment gradually became abrogated by statute in the early twentieth century and the doctrine of voluntary assumption of risk was gradually destroyed in the case decisions beginning around the late nineteenth and early twentieth centuries.

Tort remedies remain in limited form in Australia and the UK today. One argument is that the threat of tort claims is a deterrent to the employer and will encourage safer work systems. This chapter argues against this and other such claims, and indeed goes further and argues that workers' compensation systems as well, in themselves, do little or nothing to encourage prevention of workplace injury and disease.

In Australia, employees injured at work have two potential avenues of remedy through the legal system. The first is a common law action in tort (specifically negligence) or, more rarely, contract (specifically breach of implied term). The second is via one of a number of statutory workers' compensation schemes, some of which overlap, restrict or completely replace common law rights.

At common law, the tort of negligence must be proved, on the balance of probabilities, by the plaintiff (the injured worker). This requires the proof of the existence of a duty of care by the defendant (the employer), established by a legal test of reasonable forseeability, based on case precedent, or existence of an occupational health and safety statute covering the circumstances (statutory negligence) and a breach of that duty causing financially measurable damage. To receive workers' compensation benefits the injured worker must show that the injury arose out of or in the course of employment and does not need to show fault on the part of the employer. About 15% of the workforce are not covered by workers' compensation.

19.2 COMMON LAW

The ways in which the threat of a common law claim may aid in accident prevention are claimed to be as follows:

(a) by publicizing a particular accident or series of accidents that otherwise might escape public and official notice;

(b) by forcing both plaintiff employee and defendant employer to research the causes of a particular accident and the management systems within which it occurred;

(c) by imposing extra costs upon the employer, both in damages and in resisting claims, which might shift the balance of economic advantage between a particular process with its concomitant risks and alternative processes involving fewer risks and therefore claims;

(d) by penalizing the employer through the costs and other disadvantages of civil litigation, and thus marking societal disapproval of anti-social and exploitative neglect on his part or the part of someone for whom he is vicariously responsible; and

(e) by providing the victim of an accident with an outlet for feelings of outrage and resentment toward the employer, and thus providing a legal alternative to private retribution.

It is submitted that none of these arguments carry much weight. In particular, the argument (c) that civil claims would provide a financial incentive to an employer to improve safety does not hold.

Common law damages will generally be small in relation to other costs of accidents (damage to plant, disturbance of production) that the employer will bear, irrespective of the possibility of a damages claim, and in the majority of cases it will be more economic for the employer to take the risk. As the Industrial Law Society has argued:

> 'Leaving aside the effect of compulsory employers' liability insurance, the costs of accidents in most industries are so small in relation to the other costs involved in production and in accident prevention that in the majority of instances it must be more economical to take the risks and pay the compensation ... it is difficult indeed to conceive of a level of civil claims (as opposed to statutory compulsion) which would render safety investment ... economic' (Industrial Law Society, 1975).

This could change with the award of exemplary or punitive damages (d) as a means to encourage civil liability as a way of stimulating preventive activities but, some recent asbestos cases apart, the courts have shown little inclination to make such awards.

In reality, of course, employers almost always insure against common law damages claims and in many jurisdictions this is compulsory. As a result, it is rarely the person or enterprise that causes the injury or disease that pays the damage. However, the system of insurance might encourage accident prevention if the insurer varied the premiums in order to reflect the employer's accident record, greater premiums being paid by employers with higher accident frequency rates and lower premiums paid by employers with lower accident frequency rates. However, this approach has many deficiencies, which will be discussed in more detail below in relation to workers' compensation insurance.

The argument (b) is also flawed. While accident claims do fulfil a positive preventive function by drawing attention to particular hazards, causing investigation and publicity (which increase knowledge of the risks), and creating pressure for improved statutory standards, they may severely inhibit in-firm accident investigation, prevention and remedial action. Such action is inhibited by the threat of common law action because of the fear of admitting anything that may prejudice the outcome of a damages claim. Employers may be reluctant to take remedial action beyond the minimum necessary for fear that such action may be taken as an admission that pre-existing measures were inadequate.

The tort system also has an inhibitory and distorting effect on the work of making and enforcing effective regulations to prevent accidents. In parliamentary debate over drafting legislation, particular provisions are often opposed on compensation rather than accident prevention grounds. Employer/employee conflicts over civil liability implications may result in the drafting of forms of words that are compromises between the uses of law for quite different purposes. The uses of such phrases as 'reasonably practicable' to qualify the employer's duty of care tend to dilute the strength of statutory provisions [see e.g. NSW Occupational Health and Safety Act (1983) Section 15 (1) and the Victoria Industrial Safety, Health and Welfare Act (1981) Section 11 (1)].

The publicity generated by large civil claims (a) probably has stimulated preventive approaches in the case of asbestos-related cancer, dioxins and some other toxic substances, but the publicity is now being given to criminal prosecutions for manslaughter in industrial accidents.

As for the argument (e), that civil claims act as a replacement for private retribution, it should be pointed out that the objective of a court in making a damages award is to put the injured person back in his pre-accident situation, so far as money can do this, and any element of private retribution is fortuitous. It may be that an unsought side-effect of the adversary system is to foster self-righteous attitudes in the plaintiff to an excessive extent. In any event, it is difficult to see a justification for civil damages as a means of assuaging a desire for retribution that may itself have been encouraged by the whole adversarial atmosphere of the court process.

Since over 90% of civil actions are settled before hearing, it is obvious that the high cost of litigation, the difficulties of proof and the intimidatory powers of the corporate defendant deny the injured worker full compensation and hence weaken any deterrent effect. In a case that does go to hearing the ensuing battle of experts on causation or disability issues results in a system of retained specialist expert witnesses for defendants and plaintiffs, and a capacity for the legal profession to stretch out the case for the benefits of the lawyers and experts. When a settlement is reached after part hearing, costs take up a large fraction of the plaintiff's benefit. While the case is pending, there is no incentive to rehabilitation.

Furthermore, it is uncertain whether the sort of general guidance offered by the case law of negligence is much help to employers in suggesting how to avoid accidents. The law of negligence requires reasonable care to be taken according to the circumstances of the case but an employer may learn only after the result from a court whether there has been a violation. The statutory 'reasonability' clauses themselves generate a secondary common law of interpretation. The qualified nature of the general duty of care is a source of suspicion and distrust for workers, a cause of problems and uncertainty in some managements and an excuse for inertia in other employers. The qualified nature of the duty reflects the economic convenience of industry and permits industrialists to avoid some of the costs of production. It means that injuries will frequently occur under circumstances in which employers are deemed to have complied with their obligations, lending support to the view that the law is more concerned with maintaining production and profit than with ensuring safety and health at work.

19.3 WORKERS' COMPENSATION

Workers' compensation schemes were introduced supposedly to counter some of the above criticisms of the tort system. However, they still do nothing for accident prevention. The cost to the employer is still covered by insurance, the cost of premiums can be passed on to the consumer and the whole problem of accident prevention transformed into an economic rationalization. In theory, the economically rational firm, knowing that some of the costs of accidents and injuries will be charged to it through premiums, will take additional preventive measures so long as the cost of those measures is less than the anticipated benefits of reduced premiums. Therefore, it is argued, by providing employers with a financial stimulus to pay more attention to improving workplace safety, a system of differential premiums will make a positive contribution to accident and injury prevention. In practice this depends crucially on the system of premium weighting.

Broadly, there are two alternatives: a flat-rate system where every employer is charged at the same rate and one or more of three types of variable premium weighting. These can be categorized as:

(i) *classified*: premiums are adjusted to take account of variations in the risks presented by different classes of industry;

(ii) *experience rating*: premiums are varied according to the accident record of the employer (in the case of single-industry insurers like the Joint Coal Board, alternative (i) is not applicable except to distinguish open-cut from underground mining); and

(iii) *penalty rating*: an additional premium is levied against an employer with a particularly bad record, this being greater than is commensurate with the increased risk created by that record.

Neither flat nor classified rating could have any incentive effect on safety performance because in neither case is the premium rate of the individual company directly affected by changes in its accident record. However, this is not true of experience or penalty rating. Both relate premiums directly to the individual company's accident record and offer a real possibility of providing the necessary financial incentive to improve safety performance. There are, however, certain problems in practice.

Although experience rating does relate each company's safety performance directly to premium paid, it nevertheless has a number of shortcomings as a technique of accident prevention. The most serious of these is that it can only be applied satisfactorily to larger companies because only in such companies can previous accident experience be used as a reliable guide to further safety performance. In the case of smaller companies, accidents are relatively rare and the statistical basis for predicting future safety performance is too small and unstable. Atiyah (1975) claims that experience rating should only be applied to companies with over 500 employees. This would include only a very small proportion of all companies in Australia and cover only about a quarter of the total workforce. Even as applied to this minority, a scheme of experience rating is not without difficulty. Accident costs are a relatively small proportion of the total cost of manufacture of most products (<1% of wages) and any change in premium resulting from a change in accident rate would be so small as to have a negligible incentive effect. It has been argued that experience rating can probably not provide an adequate incentive to promote safety unless workers' compensation benefits are raised to reflect the full cost of accidents. Against this it should be said that in high-risk industries premiums are very high and in many companies the cost of insurance exceeds annual operating profit. In these cases, a reduction in premium of 1 or 2% could not be considered insignificant.

There is also a problem with time-lags. Calculations based on past accident records may be out of date by the time they affect premiums. This might be either because they are based on the record of a number of previous years or because the costs of serious accidents or disease are not known at the outset and are only gradually charged to the firm over future years. In either event, if the time-lag is so long that improvements in safety performance are not reflected in premiums for a number of years ahead, this will obviously not create the maximum incentive to take such action.

Employers will not react to the putative financial incentives of experience rating unless they are able to calculate the likely costs and benefits of their taking further safety precautions. If they are not able to do so, then experience-rated premiums are likely to be regarded as simply another fixed cost with little or no implication for safety performance. There are severe difficulties in making the cost-benefit calculations required. This involves estimating the probability of a certain type of accident, the likely amount of damage if it occurs, the likely reductions in insurance premiums and so on. This is the science and art of quantitative risk assessment and is a rapidly developing field. In particular, quantitative risk assessment of diseases with long latency has been practiced with some success (Nurminen *et al.*, 1992). Attempts to calculate costs and benefits of hazardous substances regulations in Australia have also been made (National Occupational Health and Safety Commission, 1993). Such exercises would be beyond the resources available to many smaller companies and the databases required are usually insufficient in size.

Finally, in calculating the experience-rated premiums, there are problems in measuring the frequency and severity of accidents. The problem of deciding whether to count one day lost time, five day lost time or no day lost time accidents and so on in the statistical basis must be dealt with. More serious is the problem of measuring accident severity. If experience rating is to fulfil the function of forcing employers to internalize the full costs of accidents, then it is these costs that must be included in calculating the premium. The difficulty is that if employers are experience rated on the basis of compensation actually paid out, then their insurance premiums will not reflect anything like the full costs of the injuries they cause. This is because workers' compensation payments only make up all or most of the wages and certain incidental expenses. Compensation for pain and suffering is limited [see e.g. Workers' Compensation Act (1987) (NSW) Section 67]. There is no compensation for indirect economic loss following an injury. Occupational diseases are grossly undercompensated and it usually requires a long struggle to have diseases prescribed as occupational.

A disturbing effect in the case of occupational disease is for the existence of potential civil liability or statutory compensation liability to mould the direction of research by scientists who are put in adversarial positions. Thus compensation of chronic bronchitis and emphysema in coal workers was delayed for many years in the UK compared to Australia. Compensation for coal workers lung disease was delayed for 30 years in the USA compared to the UK. Compensation for asbestos-related lung cancer is causing adversarial stances, resulting in severe under-recognition in the UK and Australia compared to Finland and Germany, for example. Compensation claims for dioxin-related cancer, radiation-induced cancer and bischloromethylether cancer resulted in adversarial science. A similar pattern is developing with man-made mineral fibres and non-ionizing electromagnetic radiation. RSI (occupational overuse syndrome) was a classic illustration of the clash between the psychogenic and physical aetiologic arguments. It has been argued that the class bias of physicians, engineers, scientists and even government officials in occupational health leads them to identify with management rather than the worker and hence to find ways to play down the problems of occupational disease (Bayer, 1988).

A system of penalty rating involves the imposition of an additional premium on an employer with a bad record when the additional amount of premium is greater than is commensurate with the additional risk. Ontario, Canada, provides a very good example of a system of penalty rating. Under the Ontario system, a penalty may be imposed on an employer only where three criteria are satisfied: first, where payments over the lifetime of the insurance by the Ontario Workmen's Compensation Board to injured workers exceed the premiums paid by the employer, second where the payments have exceeded the premiums in two of the preceding three years, and third where during two of the preceding three years the accident rate for the employer has been at least 25% higher than the industry norm. One feature of this system is that heavy penalties are imposed on companies with poor records but if such a company agrees to take remedial action as directed by the Board, and does so, the penalty will not be collected. The Board is also willing to exercise discretion if the employer offers a reasonable explanation for its poor record. Otherwise, a company that satisfies the three pre-conditions will be charged an extra 100% of their premium in the first penalty year and a further 25% each year thereafter, until the first pre-condition is no longer satisfied or until a ceiling of 200% is reached (Gunningham, 1984).

The philosophy behind this scheme is that employers cannot be induced to do much about safety under a system of rewards and punishments unless the punishments are sufficiently heavy to make an impact. In this respect, penalty rating has a strong advantage over experience rating, unless the latter is so structured that the full costs of accidents are reflected in the premium. An analysis of the accident records of companies penalized under the Ontario system suggests that the scheme does have a substantial effect in reducing accidents. Both the average number of accidents and the cost of accidents decreased substantially after a penalty.

The evidence of the Ontario scheme suggests that employers do respond to the financial incentives offered by penalty rating and are capable of at least a rough cost-benefit analysis that induces them to make more effort to reduce work hazards. There are, nevertheless, a number of serious limitations on the use of any system of differential premium weighting, including penalty weighting, as a technique of accident prevention. In particular, there is the conflict between the preventive and insurance objectives of workers' compensation. As an insurance programme, the aim of workers' compensation is to spread the risks of injury losses and so reduce the variability of employers' premiums. However, in fulfilling the preventive function premiums must be varied in order to reflect the costs of accidents and disease and create the necessary incentives for hazard control. This conflict can only be resolved by a trade-off between the two functions. There is little dispute that the insurance function of workers' compensation is socially valuable and must be preserved so that any compromise, such as a system of differential premiums, must retain a substantial insurance aspect. To that extent, firms will be cushioned against the full consequences of their behaviour (i.e. the cost of accidents) by the loss-spreading role of insurance and will have less than full incentive to prevent accidents. Thus, a system of differential premiums, whilst encouraging safety, can never be a fully effective technique of accident prevention.

There are also problems involved in the whole strategy of using economic incentives to encourage safety, the most serious of which is the question raised earlier of whether firms really do act in the economically rational fashion assumed in the economists' models. To the extent that they do not, economic incentives will not bring about the predicted improvement in safety performance. Some firms, rather than reducing injury and illness, may find they can reduce their insurance premiums more cheaply by playing down the health hazards to which they are exposing their workers. This strategy may be very effective since the symptoms of occupational illnesses often take years to become apparent and such diseases are notoriously difficult to trace back to their workplace origins. Proposals for registers of exposed workers have met with resistance in Australia compared to Germany or Scandinavia. The net result will be that employers successfully 'externalize' the financial costs of occupational disease onto workers, their families and society at large.

Finally, there are severe difficulties in applying a strategy of differential premiums to reduce the incidence of occupational disease. First, there are the problems referred to in the previous paragraph. Second, there is the inherent difficulty of assigning to the 'right' employer the costs of occupational diseases that do not become manifest until many years, and several employers, after a worker's exposure to a hazard. Third, there are immense problems in trying to assign costs in cases where illness is of multiple or uncertain etiology and where non-occupational factors may play a contributory role [e.g. smoking in coal workers' bronchitis and emphysema (Leigh *et al.*, 1994] and smoking in asbestos-related lung cancer (Henderson *et al.*, 1995)]. Fourth, firms face a discounting problem. They are unable to predict the likely future harm caused by occupational disease or the future benefits from a hazard-control policy so they are likely to 'discount' such future costs and to concentrate their energies and spending on more short-term objectives. Expenditure on current prevention techniques to prevent diseases such as noise-induced hearing loss and occupational cancer 20 to 40 years hence, at the risk of reducing current profits, is not an economic incentive as the present value of a liability 30 years in the future is small. In consequence of these difficulties, differential premiums are never likely to succeed in forcing internalization of the full costs of occupational disease and any contribution they do make must, at best, be of a very limited nature.

What often happens in practice is not injury and disease prevention but use of claims reduction techniques, some more legal than others. Some of these are outlined by Hopkins (Hopkins, 1995, and Chapter 20, this volume). They include pre-employment screening-out of high-risk individuals, getting injured people back to work on restricted duties, using paid sick leave, threatening dismissal if claims are made, manipulating the definition of lost-time claims for statistical purposes, deliberate minor initial claims encouraged by employers so that second more costly real claims would not count in the statistics used for premium weighting, understating payrolls and others. Commercial firms now provide expensive advice on 'how to terminate people on compensation' and other such matters.

Claims can also be contested in the courts and similar problems as for common law arise in relation to expert opinion. Indeed a large amount of secondary common law of statutory interpretation has developed. A huge medico-legal workers' compensation industry has built up in such a way that medico-legal and administrative costs at least equal the benefits to the injured worker (Creighton et al., 1983). Legal costs alone amount to 20% of the benefit. The standard benefits are relatively low but happily not so low as in parts of the USA where some economists (e.g. Viscusi, cited in McGarity and Shapiro, 1993) have advised against increasing benefits because this will encourage employees to have more accidents: 'As the degree of coverage of their income loss from job injuries is increased, workers have less of a financial incentive to avoid the injury since the size of the loss has been reduced' (McGarity and Shapiro, 1993). The viability of this industry depends on accidents occurring and disagreement between parties as to rights and liabilities, and does little or nothing for prevention.

It can also be argued that the insurance/compensation system does not assist in rehabilitation after injury. If an insurance company determines premiums according to past loss experience, and compensation payments for any one year do not exceed the amount for which provision has been made, including reinsurance, there is little incentive for an insurer to pursue doubtful claims and encourage active rehabilitation, or to encourage employers to take on partially disabled workers. This could result in the company acquiring a reputation of being a 'hard' company and potentially losing market share.

It is true that insurers and compensation authorities do invest money in accident prevention research but the processes they administer in themselves are in general counter-productive to prevention of occupational injury. Safety consciousness in employers must be motivated by other means, for example a 'moral imperative' or criminal sanctions. The first UK custodial sentence for breach of the Asbestos (Licensing) Regulations (1983) was handed down in January 1996 (British Asbestos Newsletter, 1996). Although there have been Australian prison sentences for breach of environmental regulations, we await the first such sentence for breach of occupational health and safety regulations.

References

ATIYAH, P.S. (1975) Accident prevention and variable premium rates for work-connected accidents. *Industrial Law Journal*, **4**, 1–11.
BAYER, R. (ed.) (1988) *The health and safety of workers*, Oxford: Oxford University Press.
British Asbestos Newsletter (1996) **23**, 3.
CREIGHTON, W.B., FORD, W.J. and MITCHELL, R.J. (1983) *Labor law*, p. 885, Sydney: Law Book Co.
ELLING, R.H. (1986) *The struggle for workers' health*, p. 239, New York: Baywood.
FLEMING, J.G. (1983) *The law of torts*, 6th cdn, pp. 477–496, Sydney: Law Book Co.
GUNNINGHAM, N. (1984) *Safeguarding the worker*, p. 303, Sydney: Law Book Co.
HENDERSON, D.W., ROGGLI, V.L., SHILKIN, K.B., HAMMAR, S.P. and LEIGH, J. (1995) Is asbestosis an obligate precursor for asbestos induced lung cancer? Fiber burden and the

changing balance of evidence: A preliminary discussion document. In *Sourcebook on asbestos diseases, vol. 11*, eds G.A. Peters and B.J. Peters, pp. 97–170, Charlottesville: Michie.

HOPKINS, A. (1995) *Making safety work*, pp. 32–41, Sydney: Allen & Unwin.

INDUSTRIAL LAW SOCIETY (1975) Compensation for industrial injury. *Industrial Law Journal*, **4**, 195–217.

LEIGH, J., DRISCOLL, T., BECK, R., COLE, B., HULL, B. and YANG, J. (1994) Quantitative relationship between emphysema and lung mineral content in coal workers. *Occupational and Environmental Medicine*, **51**, 400–407.

McGARITY, T.O. and SHAPIRO, S.A. (1993) *Workers at risk*, p. 22, Connecticut: Drager.

NATIONAL OCCUPATIONAL HEALTH AND SAFETY COMMISSION (1993) *Economic Impact Analaysis for Workplace Hazardous Substances Regulations*, Canberra: Australian Government Printing Service.

NURMINEN, M., CORVALAN, C., LEIGH, J. and BAKER, G. (1992) Prediction of silicosis and lung cancer in the Australian labour force exposed to silica. *Scandinavian Journal of Work, Environment and Health*, **18**, 393–399.

CHAPTER TWENTY

Does compensation have a role in injury prevention?

ANDREW HOPKINS

20.1 INTRODUCTION

It is often assumed that the workers' compensation system has a major role to play in preventing injury to workers. The assumption is that if employers pay a workers' compensation premium that reflects their claims experience they will have an incentive to improve their OHS. There is some evidence from the US that compensation systems do function in this way, although this evidence remains contestable. There is no doubt in the case of particular firms that compensation costs sometimes function in this way. This chapter argues, however, that workers' compensation costs very frequently provide no such incentive and that the compensation system is therefore a most unreliable way of achieving worker health and safety.

There are two basic reasons for this. The first is that there are many matters of great concern from an OHS point of view that in one way or another are not subject to these incentive effects since they are beyond the reach of the compensation system. The second reason stems from an examination of just how employers respond to the compensation pressures that do exist.

20.2 BEYOND THE REACH OF COMPENSATION

20.2.1 Illness

Most obviously beyond the reach of the compensation system is the matter of diseases with long latency periods, where it is often difficult to demonstrate many years after the event that any particular employer is responsible. In many respects occupational disease is a far more serious problem than occupational injury. While about 500 workers die each year from work-related injuries in Australia, it is conservatively estimated that at least 2000 die each year from work-induced cancer alone (Industry Commission, 1995). This is not reflected in workers' compensation statistics since cancer is rarely compensated. In Western Australia, for instance, although there appear to be no cases where workers have been compensated for the asbestos-related cancer mesothelioma, cancer registry figures show that the number of cases diagnosed has been rising steadily since 1975, with 41 new cases diagnosed in 1991. By 1992 the total stood at 406 (Western Australia, 1992). The annual number of new cases is expected to peak early next century. These cases can reasonably be assumed to be the result of work-related exposure to asbestos, but because of the long onset time workers have usually retired by the time their symptoms appear and they do not generate compensation costs for their former employers.

20.2.2 Rare but catastrophic events

A second safety concern that lies beyond the reach of compensation is the matter of rare but catastrophic events, for example coal-mine explosions or toppling cranes. These may cause multiple fatalities when they occur but they do not generate compensation costs to any one employer on a regular basis and so premium incentives cannot operate to ensure safety.

In the case of coal mine explosions, most mines have never had an explosion. When an explosion does occur there are certainly costs, compensation and so on but as the years roll by these costs fade from ledger books and from memories. In these circumstances, prevention of explosions depends to a considerable extent on the activities of the inspectorates, either themselves ensuring that mines are in compliance with relevant regulations or at least ensuring that self-regulation is effective.

20.2.3 Small businesses

A third problem area concerns small businesses. Because the number of compensation claims in any one small business is negligible, premiums cannot be tailored to claims experience. Thus small businesses, which is where most Australian workers are found, escape the incentive effects of the system altogether.

20.2.4 Contracting

Where a principal contractor employs a number of subcontracting firms, as on building sites, it may well be the case that most of the workers on the site are direct employees of the subcontractors, not the principal contractor. Often in such situations it is the principal contractor who is in the best position to ensure the safety of workers on the site but compensation costs fall on the subcontractor, rather than the principal, and so cannot have the desired effect.

20.2.5 The self-employed

The self-employed are a group of workers who for the most part are not covered by workers' compensation legislation. It follows that compensation pressures have no part to play in promoting the safety of these workers. Farmers, most of whom are self-employed, are a case in point. For example, the Victorian Accident Compensation Commission recorded three tractor-related deaths from mid 1985 to mid 1990. However, over the same period, the Victorian Department of Labour was notified of 37 tractor-related deaths, representing nearly 20% of all fatalities reported to the Department (Victorian Department of Labour, 1990). Clearly the compensation system has no part to play in reducing this very substantial toll.

20.3 ALTERNATIVE RESPONSES TO COMPENSATION PRESSURES

The second major set of limitations emerges from a consideration of how employers respond to the compensation costs that they incur.

20.3.1 Claims/injury management

Compensation costs are influenced not only by the number of accidents, and hence claims, but also by the duration of claims, that is the time that people are off work. It is far more effective for an employer worried about the cost of claims to try to reduce the duration, rather than the number of claims. This can be done by attending to rehabilitation, by bringing people back to work on light or alternate duties and, above all, by scrutinizing claims to ascertain whether they really necessitate time off work. Employers who engage in such active claims management find that many matters that were previously lost-time injuries can be converted to injuries without lost time, thus reducing dramatically the costs of compensation.

None of these cost reduction strategies, it must be pointed out, has anything to do with improving safety. Putting all this another way, a rational employer concerned about compensation costs will institute a more effective claim/injury management scheme. Only when this strategy has been exhausted is there an incentive to improve OHS in order to reduce costs even further.

An example, drawn from the author's own field work will illustrate the point (for further details see Hopkins, 1995). A company had recently appointed a new and energetic OHS manager. She decided to tackle one of the company's plants that regularly recorded between 30 and 40 LTIs (lost-time injuries) a year. She felt that a culture had developed at the plant in which workers saw compensation as a form of leave to which they had a *de facto* entitlement. Many of the injuries for which doctors were giving these workers time off were discovered on closer scrutiny to be relatively minor, requiring bandaging or some other medical treatment but not time off work. She adopted a policy of challenging every claim and within two years the plant was down to six LTIs in a year. This dramatic reduction was entirely attributable to tighter claims management, which resulted in the conversion of LTIs into injuries without lost time.

Once claims were under control her attention moved to safety. The firm had had two back injuries, costing a total of $A90 000, arising out of a particular work process that involved some heavy lifting. More injuries were predictable. These matters were put to top management, who agreed to buy some scissor lifters at a cost of $A15 000 each to eliminate the problem. At this point, then, compensation pressures were beginning to yield safety improvements, but only after the number of claims had been reduced as far as possible by a tighter approach to claims management.

20.3.2 Claims suppression

Yet another response of some employers to compensation costs is to suppress claims. It is difficult to know how widespread this practice is but examples abound. In the building industry, for instance, subcontractors who know that further contracts depend on a good safety record, which means in practice a small number of claims, will find ways of dissuading injured workers from claiming. They may do this by encouraging an injured worker to take sick leave rather than make a claim or, more reprehensibly, they may threaten to sack workers who make claims.

The author has come across a building industry subcontractor who at times employs up to a hundred workers but who has had only one compensation claim in nine years. This claim was made by a man who broke his shoulder and was off work for six months. However, this was by no means the only injury that employees suffered in this period. One reason why these injuries did not generate compensation claims was that the employer makes it clear that he regards compensation claimants as malingerers and bludgers. Workers who have to take time off as a result of injuries are asked to use their sick leave to cover their absence. This is clearly an illegitimate request and quite contrary to the interests of employees who, if the injury flairs at some later stage, will have no evidence that it is work-related and thus little likelihood of getting compensation. Moreover, this employer explicitly threatens to sack workers who even look as if they have the potential to make a claim. During the author's research, four workers suffered vibration injuries from long hours of work with a jack hammer. None made a compensation claim. One managed to continue working, two took sick leave and a fourth, who the employer said was malingering, was sacked. Part of the power the employer exercises over these employees stems from the temporary and uncertain nature of their employment. All knew that the current job was winding down and that some of them would be retrenched. They knew, too, that anyone who had suffered an injury would very likely be the first to go. It is obvious, then, that this employer is very deliberately and effectively suppressing legitimate compensation claims and that this is a considered response to compensation costs, a response that patently has nothing to do with improving OHS.

Even where employers do not actually threaten their employees in this way, in some situations workers believe that their jobs are at risk if they make claims and this may result in the systematic claims suppression. In one clothing factory visited by the author, when asked whether there was any under-reporting of injuries, the company nurse said that more than a dozen women had come to her within the last six months wanting treatment, mainly for repetition injuries. These workers were terrified of the consequences of coming forward in this way and implored the nurse to tell no one. She massaged their arms and gave them pain killers but felt unable to take the matter up with management. As she put it, 'I tear my hair out about what to do, but my hands are tied because of their wish not to report'. In one case the nurse insisted that a woman in severe pain see the company doctor but although an appointment was arranged the woman would not tell the doctor her story. Here, then, is a situation in which injured women work on in pain and legitimate compensation claims are suppressed.

20.3.3 Claims contestation

A final response of employers to rising claims is sometimes to challenge the legitimacy of the claims. This is particularly likely where signs are not visible to others or where there may be doubts about whether the condition is work related, as with many lung conditions. In the case of RSI, one response by employers and their insurers is to contest claims on the grounds that RSI is not a medically known condition, or that if it is not work-related. In one celebrated case in 1987 the federal taxation department won a compensation case against an employee on the grounds that since RSI was unknown by that name overseas it could not be a real medical condition (Campbell, 1988; Reid and Reynolds, 1990).

20.3.4 Organizationally structured immunity to compensation pressures

Discussion so far has concerned the various ways in which employers respond to compensation pressures, only some of which have anything to do with improving safety. However, there are also circumstances that can render managers quite immune to these pressures, ruling out even the possibility of compensation costs generating safety improvements. The problem arises in large organizations and stems from the way in which compensation costs are distributed throughout the organization. Suppose that the insurer charges the company an annual premium based on the cost of claims lodged. If corporate headquarters pays this premium direct from its own resources the various business units or budget or profit centres within the organization will not be affected by this cost and business unit managers and their subordinate line managers will have no financial incentive to do anything about claims in their own sphere of influence. What often happens in practice is that corporate headquarters does not pay the premium from its own resources but distributes the costs to the various budget centres on the basis of the number of employees working in each such centre, not on the basis of the cost of claims originating from that centre. Again, therefore, managers at these levels have no incentive to reduce the cost of claims. This means that no matter how well designed the premium scheme is from the point of view of the compensation authorities, the incentive effects are nullified by the company's internal accounting procedures.

One quite striking case from the author's research concerned a large university. Most of the budget centres within this institution had negligible numbers of claims but one section, responsible for doing maintenance work around the campus, had substantial numbers of injuries and an injury frequency rate many times higher than the campus average. Compensation costs, however, were distributed to budget centres on the basis of numbers of employees and not the cost of claims, and so this section experienced no financial pressure to improve its safety performance or even to engage in better claims management. The university's health and safety officer had targeted the maintenance section for special attention but clearly the compensation system was not providing him with any financial leverage in his campaign.

To summarize, rational employers respond to compensation costs in numerous ways, many of which have nothing to do with improving OHS. Moreover, as was shown earlier, there are many areas of work and types of employer that are beyond the reach of the compensation system altogether and where compensation costs cannot even in principle produce health and safety dividends. All of this makes the compensation system thoroughly unreliable as a means of encouraging employers to focus on injury prevention, let alone to attend to questions of health. There are ways in which the compensation system can be fine-tuned to improve its impact in this respect, but it cannot be relied on to carry the burden of prevention.

20.4 THE IMPORTANCE OF INSPECTORATES

There remains, then, a vital role for government regulations and inspectorates in encouraging and compelling employers to attend to the welfare of their workers. While it is a little beyond the scope of this chapter, some mention is made here of some of the evidence of the effectiveness of government regulatory endeavours in this respect.

Some of the strongest evidence comes from the US. Boden (1985) analyzed both fatality and disabling injury rates across all US coal mines in the period 1973–75 and found that when other relevant factors were controlled both of these indicators were inversely related to the frequency of inspection. The author concluded that 'increasing inspections by 25% would have produced a 13% decline in fatal accidents and an 18% decline in disabling accidents'.

A second piece of evidence concerns the impact of the resident inspector programme introduced by the US Mine Safety and Health Administration in the mid 1970s but abandonned in the era of deregulation initiated by the Reagan presidency. Under this programme inspectors were stationed full-time at coal mines with bad safety records. The fatality rates in these mines fell almost immediately to well below the national average (Braithwaite, 1985). The impact of the inspectorate could hardly be more graphically demonstrated.

Third, a study of the impact of the US Occupational Safety and Health Administration inspectorate showed that plants that were inspected and penalized (that is, cited for violations and ordered to pay an administrative fine) experienced a 22% decline in injuries over the following three years (Gray and Scholz, 1993). This is a very substantial decline, suggesting that OSHA is far more effective than previous studies had suggested.

Finally, some very impressive evidence emerged from a programme run by the NSW inspectorate (Young and Campbell, 1989). The programme focused on health and safety in the cotton industry in northern New South Wales. It was triggered by the high level of notifiable accidents among the workers in cotton gins: the factories that transform freshly harvested cotton into bales of lint cotton. The programme began in 1987 and ran for two years. Inspectors sought the cooperation of employers and workers and did an initial safety audit of all gins, giving advice and in some cases issuing orders. They drew up an action plan with employers that included dust control measures, machine guarding, safer work systems and improved training. Following the initial audit, they visited each gin once a month. As a result the major employer began spending about a million dollars a year on safety measures and employed a full-time safety officer. The programme was an outstanding success, reducing the annual number of accidents by 80% and the accident rate by nearly 90%. It should be noted that the figures refer to notifiable accidents (essentially serious accidents) and not workers' compensation claims. Thus improvements cannot be due simply to better claims management practices of the type discussed

Table 20.1 Accident rates for cotton gin workers in northern NSW

Year	Total no. of accidents	Total no. of operators	Accidents per operator per year
1985	40	230	0.174
1986	49	240	0.204
1987	22	260	0.083
1988	8	320	0.025

Source: Young and Campbell, 1989.

above. The study shows, in other words, that intensive industry-specific campaigns by OHS inspectorates can be dramatically effective. The evidence is presented in Table 20.1.

20.5 CONCLUSION

It is undeniable that compensation costs can exert pressures on employers to ensure the safety of their workers in some circumstances. However, it would be foolish to rely on the compensation system as the principal source of leverage in this respect. This chapter has shown that there are many ways in which employers respond to compensation costs that have nothing to do with the prevention of illness or injury. It has shown, too, that there are many circumstances where compensation pressures do not operate at all. The authorities must therefore find other ways of exerting pressure on employers to attend to worker health and safety. The role of inspectorates is vital in this respect. There is good evidence of the impact of inspectorates in reducing rates of injury, quite enough to suggest that increasing the resources available to inspectorates would be a sensible way to improve worker health and safety. Let us not embrace too enthusiastically the currently fashionable view that a well-designed compensation scheme is the best way to achieve this end.

References

BODEN, L. (1985) Government regulation of occupational safety: underground coal mine accidents 1973–75. *American Journal of Public Health*, **75**(5), 497–501.

BRAITHWAITE, J. (1985) *To punish or persuade: enforcement of coal mine safety*, Albany: State University of New York.

CAMPBELL, S. (1988) Taxing work: Susan Cooper's RSI case. Paper presented at the OHS conference, Adelaide.

GRAY, W. and SCHOLZ, J. (1993) Does regulatory enforcement work? A panel analysis of OSHA enforcement. *Law and Society Review*, **27**(1), 177–213.

HOPKINS, A. (1995) *Making safety work: getting management commitment to occupational health and safety*, Sydney: Allen and Unwin.

INDUSTRY COMMISSION (1995) *Work, health and safety*, Canberra: Australian Government Publishing Service.

REID, J. and REYROLDS, L. (1990) Requiem for RSI: the explanation and control of an occupational epidemic. *Medical Anthropology Quarterly*, **4**(2), 162–190.

VICTORIAN DEPARTMENT OF LABOUR (1990) *Workplace fatalities 1985–1990*, Melbourne: Victorian Department of Labour.

WESTERN AUSTRALIA (1992) *State of the work environment: occupational diseases*, Perth: Department of Occupational Health, Safety and Welfare.

YOUNG, R. and CAMPBELL, S. (1989) Improving health and safety in the cotton industry: employers and inspectors join forces. *Journal of Occupational Health and Safety, Australia and New Zealand*, **5**(2), 129–134.

Index

For Product Safety Concerns and Information please contact our EU
representative GPSR@taylorandfrancis.com
Taylor & Francis Verlag GmbH, Kaufingerstraße 24, 80331 München, Germany

www.ingramcontent.com/pod-product-compliance
Ingram Content Group UK Ltd.
Pitfield, Milton Keynes, MK11 3LW, UK
UKHW051828180425
457613UK00007B/249